Ernst Schering Foundation Symposium
Proceedings 2006-1
Tissue-Specific Estrogen Action

Ernst Schering Foundation Symposium
Proceedings 2006-1

Tissue-Specific Estrogen Action

Novel Mechanisms, Novel Ligands,
Novel Therapies

K.S. Korach,
T. Wintermantel
Editors

With 42 Figures

Series Editors: G. Stock and M. Lessl

Library of Congress Control Number: 2007921596
ISSN 0947-6075
ISBN 978-3-540-49547-5 Springer Berlin Heidelberg New York

This work is subject to copyright. All rights are reserved, whether the whole or part of the material is concerned, specifically the rights of translation, reprinting, reuse of illustrations, recitation, broadcasting, reproduction on microfilms or in any other way, and storage in data banks. Duplication of this publication or parts thereof is permitted only under the provisions of the German Copyright Law of September 9, 1965, in its current version, and permission for use must always be obtained from Springer-Verlag. Violations are liable for prosecution under the German Copyright Law.

Springer is a part of Springer Science+Business Media
springer.com

© Springer-Verlag Berlin Heidelberg 2007

The use of general descriptive names, registered names, trademarks, etc. in this publication does not emply, even in the absence of a specific statemant, that such names are exempt from the relevant protective laws and regulations and therefor free for general use. Product liability: The publisher cannot guarantee the accuracy any information about dosage and application contained in this book. In every induvidual case the user must check such information by consulting the relevant literature.

Editor: Dr. Ute Heilmann, Heidelberg
Desk Editor: Wilma McHugh, Heidelberg
Production Editor: Anne Strohbach, Leipzig
Cover design: WMXDesign, Heidelberg
Typesetting and production: LE-TEX Jelonek, Schmidt & Vöckler GbR, Leipzig
21/3180/YL – 5 4 3 2 1 0   Printed on acid-free paper

Preface

Nuclear hormone receptors are not only important drug targets, but have also been the focus of decades of active and highly insightful research. Ten years ago, a review on nuclear receptors was entitled "The Second Decade" and a special issue of Molecular Endocrinology in 2005 dealt with the results of these research efforts. The consensus from nuclear receptor research was of course that the signaling pathways mediated by these receptors warrant further research, even though in principle they appeared to represent the most immediate, seemingly simple signaling pathway from hormone (ligand) binding to gene expression changes.

In nuclear receptor molecular biology, estrogen receptor research has additional unique facets: since the discovery of ethinyl estradiol by Inhoffen and Hohlweg in the laboratories of Schering AG in the 1930s—and therefore several decades longer than nuclear receptor research itself—estrogen receptors have been targets of widely used, orally administered drugs. Thus, accumulating clinical experience on estrogen action in vivo helps to support the progress in molecular biological research.

Strikingly, research challenges were faced then and again when clinical observations suggested more complex effects and actions than were evident in cell culture and animal studies. On the other hand, the apparent conflicts of clinical data with well-established paradigms in preclinical models of estrogen action, e.g., in the cardiovascular system, hold both inspiration and demand for estrogen receptor research to develop better models and provide more detailed explanations of estrogen action in vivo. Newer models and explanations should not only accommodate current clinical experience, but should also open up the opportunity to develop novel treatment paradigms and better estrogen-receptor-based therapies to meet medical needs in oncology, hormone therapy, fertility control, and gynecological diseases.

In order to discuss the latest research development in this field and to foster the interaction between basic researchers and drug developers, the Ernst Schering Research Foundation held a symposium entitled "Tissue-Specific Estrogen Action: Novel Mechanisms, Novel Ligands, Novel Therapies?" The present volume of the *Ernst Schering Research Foundation Symposium Proceedings* series covers the different areas of estrogen receptor research that were the focus of the symposium: the basic molecular biology of transcriptional regulation by estrogen receptors has become dynamic thanks to the work of Frank Gannon and co-workers, who established in a series of time-resolved promoter studies that ER works in a cyclical manner on target gene promoters. The molecular biology of estrogen receptor action in vivo in both classical (reproductive) and nonclassical (e.g., blood vessels, heart, brain) target organs has been studied thoroughly using the estrogen receptor knock-out (ERKO) mouse models, as well as tissue-specific mutants (Kenneth S. Korach, Tim M. Wintermantel, and Günther Schütz). Evan Simpson elucidated estrogen actions in nonclassical target organs using both genetic models and clinical examples of estrogen deficiency. Using in vivo models for complex disease processes in the cardiovascular system, Jean-Francois Arnal studied ERKO models and carried the understanding of estrogen—and selective ER action—in atherosclerosis and vascular disease to unforeseen detail. In addition to receptor isoform-specific knockouts, receptor isoform-specific ligands gain tremendous impact on research: Theo Pelzer, Richard Williams, and Gail Risbridger used isoform-selective ligands to unravel estrogen receptor function in in vivo models of blood pressure regulation, rheumatoid arthritis, and benign prostate hyperplasia, respectively. Receptor subtype-specific ligands also hold promise in clinical development, as illustrated by Heather Harris's work on Wyeth's ERB-041 compound. Novel signaling mechanisms of estrogens such as rapid, membrane-initiated signaling are a starting point to identify mechanism-specific (as opposed to isoform-specific) ligands: Christiane Otto (Bayer Schering Pharma AG) presented 'status reports' on this new quest for ligands. Those ligands that trigger selective ER pathways might also one day provide novel treatment options, but they will without doubt generate more data to discuss in future symposia.

K.S. Korach
T.M. Wintermantel

Contents

Interfering with the Dynamics
of Estrogen Receptor-Regulated Transcription
S.A. Johnsen, S. Kangaspeska, G. Reid, F. Gannon 1

Actions of Estrogen and Estrogen Receptors
in Nonclassical Target Tissues
E. Murphy, K.S. Korach 13

Genetic Dissection of Estrogen Receptor Signaling In Vivo
T.M. Wintermantel, J. Elzer, A.E. Herbison, K.-H. Fritzemeier,
G. Schütz 25

Of Mice and Men: The Many Guises of Estrogens
E.R. Simpson, M.E. Jones 45

Estradiol Action in Atherosclerosis and Reendothelialization
J.-F. Arnal, H. Laurell, F. Lenfant, V. Douin-Echinard,
L. Brouchet, P. Gourdy 69

Functional Effects and Molecular Mechanisms
of Subtype-Selective ERα and ERβ Agonists
in the Cardiovascular System
P.A. Arias-Loza, V. Jazbutyte, K.-H. Fritzemeier,
C. Hegele-Hartung, L. Neyses, G. Ertl, T. Pelzer 87

Pathogenesis and Therapy of Rheumatoid Arthritis
R.O. Williams 107

The Role of ERα and ERβ in the Prostate: Insights
from Genetic Models and Isoform-Selective Ligands
S.J. McPherson, S.J. Ellem, V. Patchev, K.H. Fritzemeier,
G.P. Risbridger 131

Preclinical Characterization of Selective Estrogen Receptor
β Agonists: New Insights into Their Therapeutic Potential
H.A. Harris 149

Exploiting Nongenomic Estrogen Receptor-Mediated Signaling
for the Development of Pathway-Selective Estrogen
Receptor Ligands
C. Otto, S. Wessler, K.-H. Fritzemeier 163

Previous Volumes Published in This Series 183

List of Editors and Contributors

Editors

Korach, K.S.
Environmental Disease Medicine Program, NIEHS/NIH,
Research Triangle Parc, North Carolina 27709, USA

Wintermantel, T.M.
Therapeutic Research Group Gynecology and Andrology,
Female Health Care Research, Bayer Schering Pharma AG,
Müllerstraße 178, 13342 Berlin, Germany
(e-mail: tim.wintermantel@bayerhealthcare.com)

Contributors

Arias-Loza, P.A.
Medizinische Klinik I, University of Würzburg,
Josef-Schneider-Straße 2, 97080 Würzburg, Germany

Arnal, J.-F.
INSERM U589, Institut L Bugnard, CHU Rangueil,
1 Avenue Jean Poulhès, 31403 Toulouse Cedex, France
(e-mail: arnal@toulouse.inserm.fr)

Brouchet, L.
INSERM U589, Institut L Bugnard, CHU Rangueil,
1 Avenue Jean Poulhès, 31403 Toulouse Cedex, France

Douin-Echinard V.
INSERM U589, Institut L Bugnard, CHU Rangueil,
1 Avenue Jean Poulhès, 31403 Toulouse Cedex, France

Ellem, S.J.
Centre for Urology Research, Monash Institute of Medical Research,
Monash University, 27-31 Wright Street Clayton, Victoria,
Australia 3168
(e-mail: Stuart.Ellem@med.monash.edu.au)

Elzer, J.
Deutsches Krebsforschungszentrum (DKFZ),
Molecular Biology of the Cell I,
Im Neuenheimer Feld 280, 69120 Heidelberg, Germany

Ertl, G.
Medizinische Klinik I, University of Würzburg,
Josef-Schneider-Straße 2, 97080 Würzburg, Germany

Fritzemeier, K.-H.
Gynecology and Andrology, Bayer Schering Pharma AG, Berlin,
Müllerstraße 178, 13342 Berlin, Germany
(e-mail: Karlheinrich.fritzemeier@bayerhealthcare.com)

Gannon, F.
European Molecular Biology Laboratory (EMBL),
Meyerhofstraße 1, 69117 Heidelberg, Germany
(e-mail: Frank.Gannon@embo.org)

Goudry, P.
INSERM U589, Institut L Bugnard, CHU Rangueil,
1 Avenue Jean Poulhès, 31403 Toulouse Cedex, France

Harris, H.A.
Wyeth Research, Women's Health and Musculoskeletal Biology,
500 Arcola Rd, RN-3256, Collegeville PA 19426, USA
(e-mail: harrish@wyeth.com)

List of Editors and Contributors

Hegele-Hartung, Ch.
Bayer Schering Pharma AG, Müllerstraße 178, 13442 Berlin, Germany
(e-mail: Christa.Hegele-Hartung@bayerhealthcare.com)

Herbison, A.E.
Centre for Neuroendocrinology and Department of Physiology,
University of Otago School of Medical Sciences, Dunedin 9001,
New Zealand

Jazbutyte, V.
Medizinische Klinik I, University of Würzburg,
Josef-Schneider-Straße 2, 97080 Würzburg, Germany

Johnsen, S.A.
European Molecular Biology Laboratory (EMBL),
Meyerhofstraße 1, 69117 Heidelberg, Germany
(e-mail: Steven.Johnsen@embl.de)

Jones, M.E.E.
Prince Henry's Institute of Medical Research, P.O. Box 5152,
VIS 3168 Clayton, Australia

Kangaspeska, S.
European Molecular Biology Laboratory (EMBL),
Meyerhofstraße 1, 69117 Heidelberg, Germany

Laurell H.
INSERM U589, Institut L Bugnard, CHU Rangueil,
1 Avenue Jean Poulhès, 31403 Toulouse Cedex, France

Lenfant F.
INSERM U589, Institut L Bugnard,
1 Avenue Jean Poulhès, 31403 Toulouse Cedex, France

McPherson, S.J.
Centre for Urology Research, Monash Institute of Medical Research,
Monash University, 27-31 Wright Street Clayton, Victoria,
Australia 3168

Neyses, L.
Division of Cardiology, Manchester Royal Infirmary,
University of Manchester, Oxford Road, Manchester M13 9WL, UK
(e-mail: Ludwig.neyses@cmmc.nhs.uk)

Otto, Ch.
Gynecology and Andrology, Bayer Schering Pharma AG, Berlin,
Müllerstraße 178, 13342 Berlin, Germany
(e-mail: christiane.otto@bayerhealthcare.com)

Patchev, V.
Gynecology and Andrology, Bayer Schering Pharma AG, Berlin,
Müllerstraße 178, 13342 Berlin, Germany
(e-mail: Vladimir.Patchev@bayerhealthcare.com)

Pelzer, Th.
Medizinische Klinik I, University of Würzburg,
Josef-Schneider-Straße 2, 97080 Würzburg, Germany
(e-mail: pelzer_t@klinik.uni-wuerzburg.de)

Reid, G.
European Molecular Biology Laboratory (EMBL),
Meyerhofstraße 1, 69117 Heidelberg, Germany

Risbridger, G.P.
Centre for Urology Research, Monash Institute of Medical Research,
Monash University, 27-31 Wright Street Clayton, Victoria,
Australia 3168
(e-mail: gail.risbridger@med.monash.edu.au)

List of Editors and Contributors

Schütz, G.
Deutsches Krebsforschungszentrum (DKFZ),
Molecular Biology of the Cell I,
Im Neuenheimer Feld 280, 69120 Heidelberg, Germany

Simpson, E.R.
Prince Henry's Institute of Medical Research, P.O. Box 5152,
VIC 3168 Clayton, Australia
(e-mail: evan.simpson@princehenrys.org)

Wessler, S.
Paul Ehrlich Institut, Paul-Ehrlich-Straße 51–59,
63225 Langen, Germany

Williams, R.O.
Kennedy Institute of Rheumatology Division, Imperial College London,
1 Aspenlea Road, London W6 8LH, Uk
(e-mail: Richard.o.williams@imperial.ac.uk)

# Interfering with the Dynamics of Estrogen Receptor-Regulated Transcription

S.A. Johnsen(✉), S. Kangaspeska, G. Reid, F. Gannon(✉)

European Molecular Biology Laboratory (EMBL), Meyerhofstraße 1, 69117 Heidelberg, Germany
email: *Steven.Johnsen@embl.de, Frank.Gannon@embo.org*

1	Inhibiting the Proteasome	3
2	Inhibiting Specific Ubiquitin Ligases	4
3	HDAC Inhibition	5
4	Inhibition of DNA Methylation	7
5	Inhibiting RNA Polymerase II	8
6	Nontargeted Inhibition	8
7	Conclusions	9
References		10

Abstract. In recent years, there has been a growing realization that a static two-dimensional model of gene activation by transcription factors is inadequate. Based on the work from a number of groups (Kang et al. 2002; Liu and Bagchi 2004; Metivier et al. 2003; Park et al. 2005; Reid et al. 2003; Shang et al. 2000; Sharma and Fondell 2002; Vaisanen et al. 2005), it is becoming clear that transcriptional regulation by nuclear receptors is a dynamic and cyclical process (Metivier et al. 2006). There are significant consequences that arise from this shift in understanding, from nuclear receptors as ligand activated factors that bind to a response element to activate expression of a target gene to a process where the receptor repeatedly binds in order to achieve transcription. New insights that arise from viewing the activation process as cyclical and the consequences of this for developing new strategies that modulate the activity of the estrogen receptor are outlined in this chapter.

As we and others have described, the binding of the estrogen receptor-α (ERα) to an estrogen response element within the promoter of a target gene is the first of many steps that ultimately lead to the engagement and functional activation of RNA polymerase II (PolII) (Liu and Bagchi 2004; Metivier et al. 2003; Park et al. 2005; Reid et al. 2003; Shang et al. 2000). The first cellular response is the binding of SWI/SWF complex that alters the chromatin context. This is followed by modifications to the histone code by histone methyl transferase (HMT) and histone acetyl transferase (HAT)-mediated post-translational modifications to local histones at the promoter region. These changes are multiple and complex and occur in a defined sequence on a given promoter. The outcome of these alterations is the generation of a new chromatin landscape that is permissive to the recruitment of the transcription complex, mediators, elongation factors, and ultimately PolII. This rendition is not much different than the previous description of the mechanism of action of the ERα except that the cyclical process requires that the complexes assembled during the first phase need to be disassembled in order to allow for subsequent activation cycles. Consistent with this model, ERα is a target of the ubiquitin-proteasome pathway (Nawaz et al. 1999; Preisler-Mashek et al. 2002; Reid et al. 2003; Stenoien et al. 2001). Following functional attainment, events previously regarded as transcriptionally repressive then take place to reset the promoter, thereby allowing the commencement of the next cycle. The histone code is changed to a nonpermissive state through the deacetylation of histones by the action of histone deacetylases (HDACs), and the methylation of histones is also reversed. Chromatin condensation is then carried out by the chromatin remodeling complexes prior to ERα initiating the entire process again.

ERα-initiated cycles consist of an initial association phase followed by a subsequent clearance phase. The degradative action of the proteasome and the removal of the post-translational tags associated with an active histone environment lead to this conceptualization of the transcription process. The insight that ERα achieved transcriptional activation in a cyclical manner implies that interfering with any step in the process, including classical repressors of estrogen action, would likely block the assembly of the activation complex and subsequent recruitment of PolII and hence block the induction of gene expression. In many cases of breast cancer, where ERα is expressed and remains central to oncogenic proliferation, analogs of estrogen are used to block the proliferative function of ERα. It follows that an understanding of the consequences of other inhibitors that might act in an indirect way could have significant consequences. For example, inhibitors or activators of some of the steps in ERα-mediated transcriptional activation might lead to alterations in the activity of ERα and could give rise to new therapeutic reagents for the treatment of estrogen-dependent endocrine disorders such as ERα-positive breast cancer and postmenopausal osteoporosis.

1 Inhibiting the Proteasome

The ubiquitin-proteasome system is now well acknowledged as a major regulator of normal cellular physiology. Initially viewed as an end point for the degradation of proteins, it is now recognized to be a significant contributor to the maintenance of the status quo of a well-functioning cell and plays a particularly important role in the regulation of gene transcription (Muratani and Tansey 2003). The estrogen receptor itself is a direct target for ubiquitination (Nawaz et al. 1999; Nirmala and Thampan 1995). It was shown by the O'Malley (Lonard et al. 2000) and Mancini (Stenoien et al. 2001) groups that the widely used proteasome inhibitor, MG132, interfered with the ability of ERα to activate target gene transcription. This inhibition occurs despite an increase in the levels of ERα (through decreased degradation by the proteasome). We confirmed this observation and showed that the same outcome is achieved with another proteasome inhibitor lactacystin (Reid et al. 2003). Critically, and in keeping with the cyclical action of estrogen receptor mediated transactivation, we demonstrated that inhibiting polymerase activity prevents degradation of ERα, while conversely, inhibition of proteasome activity blocks mRNA synthesis from responsive genes.

In order to define the role of the proteasome in ERα-dependent transcriptional activation more clearly, we utilized chromatin immunoprecipitation analysis (ChIP) of the ERα target gene pS2 (Trefoil Factor 1) in the ERα-positive breast cancer cell line MCF7 as a model system. Induction of pS2, under the experimental conditions used, is solely dependent on ERα. We showed that proteins involved in the ubiquitin-proteasome pathway associate with the promoter of pS2 in ChIP assays. Furthermore, using transcriptionally synchronized cells, we showed that components of the ubiquitin-proteasome pathway, including certain proteasome components, were cyclically recruited to the pS2 promoter in an ERα-dependent fashion. It follows that these ERα-recruited ubiquitin-proteasome pathway proteins likely play an important role in disassembling the ERα transcriptional activation complex. This would allow for rapid and precise regulation of transcriptional activation by priming the promoter for subsequent rounds of transcriptional activation when sufficient ligand is available or by blocking additional activation when ligand is no longer present. Consistent with the vital role of the proteasome in ERα-dependent transcription, the proteasome inhibitor MG132 dramatically alters the profile of the cyclic binding of ERα to the pS2 promoter. Although ERα bound to the target promoter for a longer period of time in the presence of MG132, PolII was not recruited (Reid et al. 2003). The data obtained from these kinetic ChIP experiments are in keeping with the data from the inhibition of transactivation of a target gene in a cell and further strengthen the role of cyclic binding of ERα in the activation of gene expression.

The role of the proteasome in regulating ERα-dependent transcription may also have clinical relevance. The dipeptidyl boronic acid proteasome inhibitor PS-341 was recently accepted for therapeutic use under the name Valcade (Bortezomib) for the treatment of multiple myeloma. Interestingly, this drug also shows promise in both in vitro and xenograft experiments for use in the treatment of breast cancer (Teicher et al. 1999). From the above data, it would appear that proteasome inhibitors may potentially be included in the drugs that are used in conjunction with front-line treatments by antiestrogens for the treatment of ERα-positive cancers.

2 Inhibiting Specific Ubiquitin Ligases

Prior to degradation by the proteasome, target proteins are tagged by a series of ubiquitin moieties that are added by the action of a cascade of enzymes that culminate in the polyubiquitination of the substrate protein (Ciechanover 2005). The final component of this system, which also provides specificity, is the E3 ubiquitin ligase. It is estimated that the human genome contains more than 400 ubiquitin ligases. We showed that ERα was ubiquitinated and that both MDM2 and E6AP, members of two different classes of ubiquitin ligases (RING finger and HECT domain, respectively), were in fact present in complexes with ERα on the pS2 promoter (Reid et al. 2003). In a kinetic ChIP experiment, we showed that these ubiquitin ligases were recruited in synchrony and subsequent to ERα, but prior to the arrival of the proteasome components. This is consistent with the hypothesis that the degradation of ERα is an important component of the transcription cycle of ERα. In the presence of proteasome inhibitor, the E3 ligases, but not the proteasome component Rpt6 (or PolII), were still recruited to the pS2 promoter together with ERα. One would predict, therefore, that inhibitors of specific ER-associated E3 ligases should also interfere with the ability of ERα to transactivate a target gene. One such candidate is the Nutlin family of compounds developed by Hoffman-La Roche. The Nutlins are specific inhibitors of the interaction between MDM2 and p53 (Vassilev et al. 2004) and show great potential for therapeutic use in the treatment of many types of cancer. The impact of these compounds on ERα action will be very instructive and could potentially point to a further novel approach to the treatment of ERα-positive cancers.

3   HDAC Inhibition

As indicated above, an integral part of the transcription cycle for estrogen-regulated genes is the removal of acetylation tags from histones. The histone deacetylases (HDACs) are a family of enzymes responsible for this step and have been shown to be recruited to the pS2 promoter during transcriptional activation at the time when the binding phase of ERα has been completed (Metivier et al. 2003). Although deacetylation is frequently associated with the shut-down of a chromatin locus, it was critical to determine whether the inhibition of the HDAC activity indeed promoted gene expression by interfering with the shutting down of the chromatin locus or inhibited it by blocking the integrated steps in the cyclical binding of the transcription factor ERα. A medicinally relevant alternative to Trichostatin A (TSA), the reagent most commonly used to block HDAC activity in the laboratory, is valproic acid (VPA). Interestingly, VPA is widely used clinically for the treatment of convulsions, epilepsy, and related disorders and is currently in phase II trials as a potential cancer therapy (Chavez-Blanco et al. 2005). The reported side effects of VPA treatment include many that one might anticipate from an interruption of the normal physiological function of ERα such as reproductive abnormalities (Duncan 2001; Isojarvi et al. 1993; O'Donovan et al. 2002), endocrine disorders (Rattya et al. 2001), and decreased bone mass (Sato et al. 2001). Consistent with these results, we demonstrated that when cells were transfected with an estrogen-responsive reporter construct (ERE-TK-luciferase) and treated with VPA (or TSA), the transactivation was greatly diminished (Reid et al. 2005). The consequences at the cellular level were also profound, with the compounds showing a cytotoxicity profile that matched their inhibition of reporter gene activation. Using expression microarray assays, it was further shown that there was a 90% overlap between the effects of VPA and tamoxifen (the estrogen-receptor ligand used extensively in the treatment of breast cancer) on a family of estrogen-responsive genes. Taken together it is clear that the inhibition of histone deacetylase activity did not stimulate transcription of ERα activated genes, but rather inhibited the expression of these genes, thus adding further support to the proposed essential role for the integration of all steps in the cyclical induction of transcription by ERα.

Interestingly, the biological effects of VPA also appear to be linked to the ubiquitin-proteasome pathway. For example, while VPA treatment decreased ERα mRNA expression, it also decreased ERα protein levels in a proteasome-dependent manner (Reid et al. 2005). Furthermore, VPA treatment has been shown to increase targeted protein degradation by specifically increasing the expression of the ubiquitin-conjugating enzyme UBC8, thereby promoting the ubiquitination and subsequent proteasomal degradation of substrate proteins by the ubiquitin ligase RNF12 (Kramer et al. 2003). We are currently investigating whether RNF12 also plays a role in the regulation of ERα-dependent transcription and whether this may account for the proteasome-dependent function of VPA in mediating increased ERα degradation.

Arising from these studies, yet another family of therapeutic agents can be viewed as having potential in the treatment of breast cancer. As indicated above, VPA is widely used at high concentrations for other disorders and is beginning to be investigated for its therapeutic efficacy in cancer. A study is being prepared to follow the epidemiology of women that received VPA for a different dysfunction with respect to the incidence of breast cancer. In light of our results, one would predict that women who have undergone VPA treatment for other disorders would have a lower risk of breast cancer. However, our lack of understanding regarding some of the early stages of the onset of breast cancer could give rise to a different outcome. It would perhaps be surprising if there were no effect on ERα-mediated cancers and therefore the outcome of this study will be of great interest.

4   Inhibition of DNA Methylation

The mechanism of action of VPA on estrogen-regulated transcription was examined in detail by ChIP experiments using the pS2 promoter as an indicator of ERα binding and activity. These experiments suggested a role for the involvement of the maintenance DNA methyl transferase-1 (DNMT1) and methyl CpG binding protein-2 (MeCP2) in the activation of gene transcription. In the absence of the HDAC inhibitor, association of DNMT1 and MeCP2 was readily detected by ChIP analysis, whereas treatment with VPA increased binding of MeCP2, but decreased the presence of DNMT1 on the promoter (Reid et al. 2005). Furthermore, assays using the methylation-sensitive restriction enzyme HpaII as an indicator of the methylation status of CpG dinucleotides in the vicinity of the promoter demonstrated that a change in the methylation status occurred at specific CpGs near the transcriptional start site (Reid et al. 2005). Moreover, re-expression of ERα in an ERα-negative cell line MDA-MB-231 reversed the repressed state of the pS2 promoter (Metivier et al. 2004; Reid et al. 2005). All of these data suggest that there are variations in the methylation status of the actively transcribed promoter under study in MCF7 cells, and that this may be linked in some way with ER cycling and HDAC recruitment to the locus. It follows that interference with the methylation status of a promoter should also block the activation of the gene.

5 Inhibiting RNA Polymerase II

The obvious and inevitable consequence of inhibiting PolII would be the blocking of transcription. But should that have an influence on the ERα in the transactivation cycle? It could have been that there was no effect, but if ERα and PolII are components of the same transcription environment, we reasoned that there could be some feedback on the activator (ERα). In support of this, ERα protein levels decrease to approximately 10% of that in cells grown in estrogen-free medium upon the addition of estrogen. The engagement of ERα in activating transcription therefore usually results in a general destabilization of ERα. Interestingly, upon the addition of tamoxifen, ERα levels increase. It is known that tamoxifen recruits co-repressors in breast cancer cells and suggests that the stabilization may arise from a conformational change of ERα in the presence of tamoxifen or because of some steric hindrance to the degradative process by the presence of the co-repressors. When we tested the stability of a naturally occurring truncated ERα isoform (ER46) (Flouriot et al. 2000), we found that this version of ERα is stable even in the presence of E2. All of these data suggest that the regulation of ERα stability is at the center of complex interactions that appear to be related to its engagement in gene activation. For these reasons, it was not surprising that the use of inhibitors of PolII resulted in the stabilization of ERα (Reid et al. 2003). The image is therefore reinforced that ERα and PolII are the start and end points of the cycle of gene activation and that, like cogs in a wheel, any blockage in this cycle inevitably leads to effects on the other components of the system, even if their activities are required at different time points in the cycle.

6 Nontargeted Inhibition

Since the ERα transcription activation cycle was described, approximately 50 proteins with a variety of functions were shown to cycle on the pS2 target promoter in synchrony with the ERα. The total range of participants in the process, however, could be much greater that that. The inhibition studies outlined above had a specific target. We reasoned that new components of the transcription activation cycle could be identified by a screen of diverse chemical compounds. Therefore we tested a library of 55,000 compounds with drug potential for their effects on the MCF7 (ERα-positive) cell line transiently transfected with an estrogen response element reporter construct. Most compounds identified in the screen had no effect. However, approximately 1% inhibited the induction of ERα activity while 1% increased the activity. Focusing on the inhibitors, we showed that five different chemical classes were involved with multiple hits coming from closely related compounds. The inhibition of ERα action was achieved with approximately the same concentration of the compounds required to inhibit proliferation. The analysis to date suggests that none of these compounds have targets that have been analyzed for their role in ERα-mediated transcription. However, it should be noted that the effects on ERα could be indirect and that further studies on the kinetics of the inhibitors will be required. If the ERα-modulated process is not a target, but is coincidentally blocked due to destruction or misregulation of some other key steps in the physiology of the cell, it could be equally instructive, as the goal in cancer treatment is generally to destroy the tumor cells.

7 Conclusions

The primary goal of most medically relevant research is to understand the basic biology of a system so that this knowledge can then be used in a practical manner to develop new therapies. For those working on ERα, longer-term targets include breast cancer and osteoporosis. By focusing on the actions of ERα when it acts as a transcription factor, it has become clear that ERα binds to the promoter, recruits, in a sequential and programmed manner, a series of proteins that change the promoter region profoundly and prepare the local context for the engagement of PolII, thus leading to transcriptional activation. The binding of ERα and the assembly of the activation complex has been shown to be cyclic and all evidence obtained to date is concordant with a very tight interlinkage between all steps of the cycle and between the subsequent cycles. The productive process eventually stops when the level of estrogen in the system diminishes, as happens under normal physiological conditions such as during the estrous cycle and at the onset of menopause. The new insights on how this transcription factor works are of general interest, but the focus of this report is the consequence of this knowledge for treatment of cancer. We have shown that when steps in the cycle have been blocked by well-known inhibitors, the whole process was interrupted. As many of these inhibitors are used in clinical settings for diseases other than breast cancer, it follows that new mechanisms of modulating ERα activity have been identified. In addition to inhibitors of the proteasome, specific E3 ligases and HDACs, the above-described chemical screen demonstrates that ERα-mediated transcription is a potentially fruitful source of new and improved therapeutic agents. Obviously, the steps between an inhibitor and a drug are fraught with difficulties and many such approaches lead to disappointment. Nonetheless, the fact that basic research repeatedly gives rise to such hopes validates the general view that investment in the generation of knowledge is the correct approach to identifying new therapeutic compounds.

References

Chavez-Blanco A, Segura-Pacheco B, Perez-Cardenas E et al. (2005) Histone acetylation and histone deacetylase activity of magnesium valproate in tumor and peripheral blood of patients with cervical cancer. A phase I study. Mol Cancer 4:22

Ciechanover A (2005) Proteolysis: from the lysosome to ubiquitin and the proteasome. Nat Rev Mol Cell Biol 6:79–87

Duncan S (2001) Polycystic ovarian syndrome in women with epilepsy: a review. Epilepsia 42(Suppl 3):60–65

Flouriot G, Brand H, Denger S et al. (2000) Identification of a new isoform of the human estrogen receptor-alpha (hER-alpha) that is encoded by distinct transcripts and that is able to repress hER-alpha activation function 1. EMBO J 19:4688–4700

Isojarvi JI, Laatikainen TJ, Pakarinen AJ et al. (1993) Polycystic ovaries and hyperandrogenism in women taking valproate for epilepsy. N Engl J Med 329:1383–1388

Kang Z, Pirskanen A, Janne OA et al. (2002) Involvement of proteasome in the dynamic assembly of the androgen receptor transcription complex. J Biol Chem 277:48366–48371

Kramer OH, Zhu P, Ostendorff HP et al. (2003) The histone deacetylase inhibitor valproic acid selectively induces proteasomal degradation of HDAC2. EMBO J 22:3411–3420

Liu XF, Bagchi MK (2004) Recruitment of distinct chromatin-modifying complexes by tamoxifen-complexed estrogen receptor at natural target gene promoters in vivo. J Biol Chem 279:15050–15058

Lonard DM, Nawaz Z, Smith CL et al. (2000) The 26S proteasome is required for estrogen receptor-alpha and coactivator turnover and for efficient estrogen receptor-alpha transactivation. Mol Cell 5:939–948

Metivier R, Penot G, Hubner MR et al. (2003) Estrogen receptor-alpha directs ordered, cyclical, and combinatorial recruitment of cofactors on a natural target promoter. Cell 115:751–763

Metivier R, Penot G, Carmouche RP et al. (2004) Transcriptional complexes engaged by apo-estrogen receptor-alpha isoforms have divergent outcomes. EMBO J 23:3653–3666

Metivier R, Reid G, Gannon F (2006) Transcription in four dimensions: nuclear receptor-directed initiation of gene expression. EMBO Rep 7:161–167

Muratani M, Tansey WP (2003) How the ubiquitin-proteasome system controls transcription. Nat Rev Mol Cell Biol 4:192–201

Nawaz Z, Lonard DM, Dennis AP et al. (1999) Proteasome-dependent degradation of the human estrogen receptor. Proc Natl Acad Sci USA 96:1858–1862

Nirmala PB, Thampan RV (1995) Ubiquitination of the rat uterine estrogen receptor: dependence on estradiol. Biochem Biophys Res Commun 213:24–31

O'Donovan C, Kusumakar V, Graves GR et al. (2002) Menstrual abnormalities and polycystic ovary syndrome in women taking valproate for bipolar mood disorder. J Clin Psychiatry 63:322–330

Park KJ, Krishnan V, O'Malley BW et al. (2005) Formation of an IKKalpha-dependent transcription complex is required for estrogen receptor-mediated gene activation. Mol Cell 18:71–82

Preisler-Mashek MT, Solodin N, Stark BL et al. (2002) Ligand-specific regulation of proteasome-mediated proteolysis of estrogen receptor-alpha. Am J Physiol Endocrinol Metab 282:E891–E898

Rattya J, Pakarinen AJ, Knip M et al. (2001) Early hormonal changes during valproate or carbamazepine treatment: a 3-month study. Neurology 57:440–444

Reid G, Hubner MR, Metivier R et al. (2003) Cyclic, proteasome-mediated turnover of unliganded and liganded ERalpha on responsive promoters is an integral feature of estrogen signaling. Mol Cell 11:695–707

Reid G, Metivier R, Lin CY et al. (2005) Multiple mechanisms induce transcriptional silencing of a subset of genes, including oestrogen receptor alpha, in response to deacetylase inhibition by valproic acid and trichostatin A. Oncogene 24:4894–4907

Sato Y, Kondo I, Ishida S et al. (2001) Decreased bone mass and increased bone turnover with valproate therapy in adults with epilepsy. Neurology 57:445–449

Shang Y, Hu X, DiRenzo J et al. (2000) Cofactor dynamics and sufficiency in estrogen receptor-regulated transcription. Cell 103:843–852

Sharma D, Fondell JD (2002) Ordered recruitment of histone acetyltransferases and the TRAP/Mediator complex to thyroid hormone-responsive promoters in vivo. Proc Natl Acad Sci USA 99:7934–7939

Stenoien DL, Patel K, Mancini MG et al. (2001) FRAP reveals that mobility of oestrogen receptor-alpha is ligand- and proteasome-dependent. Nat Cell Biol 3:15–23

Teicher BA, Ara G, Herbst R et al. (1999) The proteasome inhibitor PS-341 in cancer therapy. Clin Cancer Res 5:2638–2645

Vaisanen S, Dunlop TW, Sinkkonen L et al. (2005) Spatio-temporal activation of chromatin on the human CYP24 gene promoter in the presence of 1alpha,25-Dihydroxyvitamin D3. J Mol Biol 350:65–77

Vassilev LT, Vu BT, Graves B et al. (2004) In vivo activation of the p53 pathway by small-molecule antagonists of MDM2. Science 303:844–848

Actions of Estrogen and Estrogen Receptors in Nonclassical Target Tissues

E. Murphy, K.S. Korach(✉)

Laboratory of Reproductive and Developmental Toxicology, National Institute of Environmental Health Sciences, National Institute of Health, 111 Alexander Drive, Research Triangle Park, 27709 North Caroline, USA
email: *korach@niehs.nih.gov*

1	Gender Differences in Ischemia–Reperfusion Injury	14
2	Gender Differences in Hypertrophy	15
3	Role of ER-α and ER-β in Cardiovascular Disease	16
3.1	Ischemia–Reperfusion Injury	16
3.2	Hypertrophy	18
4	Direct Versus Indirect Effects of Estrogen in Heart	19
5	Summary	20
References		20

Abstract. Hormonal effects on classical endocrine target organs such as the female reproductive tract, mammary gland, ovary, and neuroendocrine system have been thoroughly studied, with significant advancements in our understanding of estrogen actions and disease conditions from both cell culture as well as new experimental animal models. Knowledge of the highly appreciated effects of estrogen in nonclassical endocrine organ systems, arising from epidemiological and clinical findings in the cardiovascular, immune, GI tract, and liver, is only now becoming clarified from the development and use of knock-out or transgenic animal models for the study of both estrogen and ER activities. There are considerable epidemiological data showing that premenopausal females (Barrett-Connor 1997; Crabbe et al. 2003) have reduced risk for cardiovascular disease. However, a recent large clinical trial failed to show cardioprotection for postmenopausal females on estrogen–progestin replacement (Rossouw et al. 2002). In fact, the Women's Health Initiative Study showed increased cardiovascular risk for females taking an estrogen–progestin combination. These studies suggest that we need a better understanding of the mechanisms responsible for cardioprotection in females.

1 Gender Differences in Ischemia–Reperfusion Injury

Gender differences in cardiovascular response have been observed in some animal studies, suggesting that compared to males, intact females (without exogenous estrogen treatment) have reduced ischemia–reperfusion injury (Bae and Zhang 2005; Wang et al. 2006). Other studies show that in unstimulated, wild-type hearts, there is no male–female difference in ischemia–reperfusion injury (Przyklenk et al. 1995; Li and Kloner 1995; Cross et al. 1998). However, there are a number of studies showing that treatment of animals or perfused hearts with exogenous estrogen (typically pharmacological doses) can reduce ischemia–reperfusion injury (Li and Kloner 1995; Booth et al. 2005; Das and Sarkar 2006). Das and Sarkar (2006) reported that pretreatment of rabbits with estradiol (10 mg/kg i.v.) prior to coronary artery ligation significantly reduced infarct size (19% vs 40%). They further showed that pretreatment with 5 hydroxydecanoate (5HD; an inhibitor of the mito KATP channel) blocked the infarct size reduction afforded by estradiol. Furthermore, in a number of transgenic mouse models with increased contractility, females have reduced ischemia–reperfusion injury (Cross et al. 1998, 1999, 2003). Also under conditions of increased contractility/increased cell calcium such as occurs with addition of isoproterenol or elevated perfusate calcium, females exhibit reduced ischemia–reperfusion injury (Cross et al. 2002b; Gabel et al. 2005; Kam et al. 2004). Kam et al. (2004) showed in Langendorff perfused hearts treated with isoproterenol that hearts from intact females had smaller infarcts than hearts from ovariectomized females, which they suggested was related to a higher expression of beta1-adrenergic receptors (AR) in hearts from ovariectomized females. The lower expression of beta-1 AR in intact females would provide a mechanism for the reduced calcium overload (Chen et al. 2003) and reduced ischemia–reperfusion injury observed with isoproterenol; however, the decrease in beta1-AR in females would not easily account for the reduced injury observed in the transgenic models or wild-type hearts with increased extracellular calcium.

The reduced ischemic injury in hypercontractile females could be due to a reduction in calcium overload in females or alternatively females could have a similar calcium loading but have less injury due protection perhaps mediated by increased PI-3 kinase activity (Simoncini et al. 2000), for example. It appears, however, that hypercontractile females have less calcium loading. For example, in transgenic hearts with overexpression of plasma membrane sodium–calcium exchanger (NCX) (Sugishita et al. 2001) and wild-type hearts with addition of isoproterenol (Chen et al. 2003), females have been shown to have less calcium loading. These data suggest that estrogen reduces calcium loading in these genetic models and with isoproterenol such that there is less calcium overload at the start of ischemia, which results in less ischemic injury, since elevated calcium has clearly been shown to increase ischemia-reperfusion injury (Steenbergen et al. 1987; Murphy et al. 1991; Baines et al. 2005). The mechanism by which estrogen modulates intracellular calcium is likely to be complex, but it appears to be mediated, at least in part, by nitric oxide synthase (NOS). Estrogen is well know to up-regulate NOS (Nuedling et al. 2001; Sun et al. 2006), and inhibitors of nitric oxide synthase (NOS) block the protection in females (Cross et al. 2002a, 2003; Sun et al. 2006), suggesting a role for NO. Neudling et al. (2001) reported that COS7 cells transfected with ER-β, but not ER-α, resulted in activation of eNOS and iNOS.

2 Gender Differences in Hypertrophy

Females have also been shown to have reduced cardiac hypertrophy compared to males (Skavdahl et al. 2005) and compared to ovariectomized females (van Eickels et al. 2001). Treatment of ovariectomized females with estrogen has also been reported to reduce hypertrophy (van Eickels et al. 2001). Many animal models of heart failure have reported that females have improved survival and/or improved contractile function (Olsson et al. 2001; Kadokami et al. 2000; Dash et al. 2003). Beer et al. (2006) also found that treatment of female rats with 17β-estradiol (7.5 mg/90 days) for 2 weeks prior to and 8 weeks following myocardial infarction prevented the development of heart failure that occurred in untreated hearts. Similarly, Kadokami et al. (2005) reported that 17β-estradiol improved survival in male mice with a cardiomyopathy induced by overexpression of tumor necrosis factor-α.

3 Role of ER-α and ER-β in Cardiovascular Disease

Two estrogen receptors, ER-α and ER-β, are known to be expressed in the cardiovascular system. To identify the specific estrogen receptor involved in cardioprotection in females, studies were done using mice lacking ER-α and mice lacking ER-β (Wang et al. 2006; Gabel et al. 2005; Skavdahl et al. 2005; Pelzer et al. 2005; Babiker et al. 2006). Studies have also been conducted using ER-α- and ER-β-selective agonists (Booth et al. 2005; Yu et al. 2006; Hsieh et al. 2006).

3.1 Ischemia–Reperfusion Injury

Following brief treatment with isoproterenol, hearts were subjected to ischemia and reperfusion, and postischemic contractile function and infarct size were measured in wild-type male and female mouse hearts, and female αERKO and βERKO hearts. Wild-type males exhibited significantly poorer functional recovery and more necrosis than wild-type females. αERKO females exhibited ischemia–reperfusion injury similar to that observed in wild-type females, whereas βERKO females exhibited significantly poorer functional recovery and more necrosis than wild-type females and were more similar to wild-type males. Using a model of trauma hemorrhage shock injury, Chaudry and co-workers (Yu et al. 2006; Hsieh et al. 2006) found that following trauma–hemorrhage shock males have depressed cardiovascular function that can be reversed by administration of 17β-estradiol just following the trauma hemorrhage. Hsieh et al. (2006) found that 24 h following trauma hemorrhage there is a decrease in PGC-1α and ATP levels. They report that this decline in PCG-1α and ATP was reversed if estradiol or an ER-β selective agonist (DNP) was administered just following trauma hemorrhage. These data suggest that the beneficial effects of estrogen in trauma hemorrhage may be mediated by ER-β up-regulation of PGC-1a. In another study by this group, Yu et al. (2006) reported that trauma hemorrhage resulted in a decrease in Hsp 32, 60, 70 and 90 mRNA and HSF-1 DNA binding and that these effects were blocked by administration of an ER-β agonist (DPN). In addition, Hsieh et al. (2006) reports that mitochondrial ER-β is important for up-regulation of mitochondrial respiratory complex proteins, and that DPN administration protects in a trauma hemorrhage model by activation of mitochondrial ER-β. These data suggest that estrogen, through the β-estrogen receptor, plays a role in the protection observed in the female heart.

In contrast to these studies, Zhai et al. (2000) found that hearts from αERKO mice showed increased injury in a model in which hearts were subjected to 45 min of global ischemia at 4 °C followed by 180 min of oxygenated reperfusion at 37 °C. Male αERKO hearts started beating later and had more fibrillation than wild-type hearts. Wang et al. (Wang et al. 2006) subjected Langendorff perfused mouse hearts to 20 min of ischemia and 60 min of reperfusion and found that male hearts had poorer recovery of the rate of change in developed pressure (dP/dt). They further showed that female mice lacking ER-α had a similar recovery of dP/dt to males that was worse than that observed in wild-type females. Also consistent with a role for ER-α in different model of cardioprotection, Booth et al. (2005) reported that i.v. administration of an ER-α selective agonists, PPT (3 mg/kg) 30 min prior to coronary occlusion significantly reduced infarct size (18% PPT vs 45% vehicle) in rabbit hearts. The protection afforded by PPT was blocked by co-administration of ICI-182,780. Administration of the ER-β selective agonists DPN (3 mg/kg) did not reduce infarct size (45%). In ovariectomized rabbits, PPT at a 3-mg/kg dose also reduced infarct size, although it was not as protective as estradiol. However, 10 mg/kg of PPT was as protective as estradiol. It is not clear how the doses of PPT and DPN were chosen and how these doses related to the dose response of other target tissues (e.g., uterine weight or gonadotropin inhibition).

Thus there is no clear consensus regarding the role of ER-α vs ER-β in cardioprotection. This discrepancy may be due to the different models of injury. ER-β appears to be important in a model of global ischemia and reperfusion and a model of trauma hemorrhage, whereas ER-α appears to be important in a model of cardioplegia or when given as a bolus at high doses prior to ischemia. In addition, it is not clear whether the protective effects of estrogen are mediated by direct effects on cardiomyocytes, the vasculature, or some other target tissue cell types. Clearly, additional studies will be necessary to delineate the relative role of ERs in cardioprotection.

3.2 Hypertrophy

In contrast to the discrepancies regarding the role of ER-α and ER-β in protection from ischemia–reperfusion injury, there is good agreement that ER-β is important for the reduced hypertrophy observed in many models. Skavdahl et al. performed transverse aortic constriction (TAC) and sham operations in male and female wild type, α-ERKO, and β-ERKO mice (Skavdahl et al. 2005). Body, heart, and lung weights were measured 2 weeks after surgery. Wild-type male mice subjected to TAC showed a 64% increase in the heart to body weight (HW/BW) ratio compared to sham. Wild-type female mice subjected to TAC showed a 31% increase in HW/BW compared to sham, which was significantly less than their male counterparts. α-ERKO females developed a HW/BW ratio nearly identical to that seen in wild-type littermate females in response to TAC, indicating that ER-α is not essential for the attenuation of hypertrophy observed in wild-type females. In contrast, β-ERKO females responded to TAC with a significant increase in the HW/BW ratio compared to wild-type littermate females, indicating an important role for the ER-β in attenuating the hypertrophic response to pressure overload. Similarly, Pelzer et al. (2005) have reported that mice lacking ER-β have increased mortality and increased pro-ANP in heart failure due to myocardial infarction. Also consistent with a role for ER-β in hypertrophy, Babiker et al. (Babiker et al. 2006) used ER-α- and ER-β-deficient mice and showed that ER-β mediates the estradiol-dependent reduction in left ventricular (LV) hypertrophy following transaortic constriction. Thus, ER-β appears to be involved in the reduced hypertrophy observed in females. Interestingly, Peter et al. (2005) showed that in women, but not men, two polymorphisms in ER-β (ERS2 rs1256031 and ERS2 rs1256059) were associated with LV mass and LV wall thickness. The protective effects of estrogen with regard to cardiac hypertrophy also appear to be age-dependent. Jazbutyte et al. (2006) show that estrogen administration to young and senescent ovariectomized SHR rats inhibited uterus atrophy and gain of body weight, but cardiac hypertrophy was attenuated only in the young rats.

4 Direct Versus Indirect Effects of Estrogen in Heart

The protection observed in females and with addition of estrogen may be mediated by indirect systemic effects or by direct effects on cardiomyocytes. Jovanovic et al. (2000) showed that cardiomyocytes pretreated with estradiol (10 nM) had reduced cytosolic calcium overload following metabolic inhibition (3 min exposure to dinitrophenol followed by washout of dinitrophenol). These data suggest that estrogen has direct cardioprotective effects on cardiac myocytes. However, some investigators have questioned whether either ER is present in heart (Forster et al. 2004). Clearly, additional studies are needed to define the mechanism for the protection observed in females. Studies with cardiac-specific loss of ER-α and ER-β would help in defining the direct effects of estrogen in heart.

There is also a controversy regarding whether mitochondria contain functional estrogen receptors. Yang et al. (2004) used immunocytochemistry, immunoblotting, and mass spectrometry to show that ER-β localizes to the mitochondria. However, Schwend and Gustaffson (2006) have questioned the MALDI-TOF identification of ER-β. Others have also reported mitochondrial localization of ER. Pedram et al. (2006) find ER-α and ER-β in MCF-7 and endothelial cells. They also find that estrogen inhibits UV-induced cytochrome c release, decreased in mitochondrial membrane potential, ROS production, and apoptosis. To determine whether these estradiol effects on mitochondria are direct vs indirect via nuclear transcriptional regulation or membrane-bound ER activating PI-3 kinase (Simoncini et al. 2000), the ligand-binding domain of ER-α was targeted to the plasma membrane or the nucleus or the mitochondria in HCC-1569 or CHO cells. With nuclear localization of ER-α, addition of estrogen did not protect from UV irradiation. However, both mitochondrial and membrane localized ER-α provided protection. In another model, Pedram et al. (2006) showed that UV irradiation of mitochondrial induces cytochrome c release, which was blocked by addition of estrogen to the mitochondria. They further showed that the ER-β-selective agonist (DPN) was more potent than the ER-α-selective agonist (PPT) in inhibiting cytochrome c release, suggesting ER-β as the mediator of the action. Parkash et al. (2006) report that estrogen addition to MCF7 cells can modulate mitochondrial calcium uptake. Similarly, Lobaton et al. (2005) reported that several agonist and antagonist of estrogen receptors modulate calcium uptake into the mitochondria.

5   Summary

More definitive evidence is emerging that estrogen is producing effects on a variety of organ systems not previously thought to be directly responsive. As for the cardiovascular system, it would appear that the two ER proteins provide differential and unique functions, for ER-β in the heart, and ER-α in the peripheral vasculature. Additionally, besides the well known and studied areas of gene regulation, estrogen appears to influence cellular functions involving other actions besides gene regulation, such as Ca^{+2} signaling or mitochondrial response, which are now being more appreciated as part of its physiological action. Continued research in these directions will undoubtedly uncover even more novel actions for estrogen and the ER. Determining the specific mechanisms and ER forms mediating the effects will then provide new therapeutic approaches for expanding the development of ER-selective agonists and antagonists to produce more effective treatments.

References

Babiker FA, Lips D, Meyer R et al. (2006) Estrogen receptor beta protects the murine heart against left ventricular hypertrophy. Arterioscler Thromb Vasc Biol 26:1524–1530

Bae S, Zhang L (2005) Gender differences in cardioprotection against ischemia/reperfusion injury in adult rat hearts: focus on Akt and protein kinase C signaling. J Pharmacol Exp Ther 315:1125–1135

Baines CP, Kaiser RA, Purcell NH et al. (2005) Loss of cyclophilin D reveals a critical role for mitochondrial permeability transition in cell death. Nature 434:658–662

Barrett-Connor E (1997) Sex differences in coronary heart disease. Why are women so superior? The 1995 Ancel Keys Lecture. Circulation 95:252–264

Beer S, Reincke M, Kral M et al. (2006) High-dose 17beta-estradiol treatment prevents development of heart failure post-myocardial infarction in the rat. Basic Res Cardiol Jul 4 [Epub ahead of print]

Booth EA, Obeid NR, Lucchesi BR (2005) Activation of estrogen receptor-alpha protects the in vivo rabbit heart from ischemia-reperfusion injury. Am J Physiol Heart Circ Physiol 289:H2039–H2047

Chen J, Petrank J, Yamamura K et al. (2003) Gender differences in sarcoplasmic reticulum calcium loading after isoproterenol. Am J Physiol Heart Circ Physiol 285:H2657–H662

Crabbe DL, Dipla K, Ambati S et al. (2003) Gender differences in post-infarction hypertrophy in end-stage failing hearts. J Am Coll Cardiol 41:300–306

Cross HR, Lu L, Steenbergen C et al. (1998) Overexpression of the cardiac Na+/Ca2+ exchanger increases susceptibility to ischemia/reperfusion injury in male, but not female, transgenic mice. Circ Res 83:1215–1223

Cross HR, Steenberger C, Lefkowitz RJ et al. (1999) Overexpression of the cardiac beta-adrenergic receptor and expression of a beta-adrenergic receptor kinase-1 (betaARK1) inhibitor both increase myocardial contractility but have differential effects on susceptibility to ischemic injury. Circ Res 85:1077–1084

Cross HR, Murphy E, Koch WJ et al. (2002a) Male and female mice overexpressing the beta-adrenergic receptor exhibit differences in ischemia/reperfusion injury: role of nitric oxide. Cardiovasc Res 53:662–671

Cross HR, Murphy E, Steenbergen C (2002b) Ca(2+) loading and adrenergic stimulation reveal male/female differences in susceptibility to ischemia-reperfusion injury. Am J Physiol Heart Circ Physiol 283:H481–H489

Cross HR, Kranias EG, Murphy E et al. (2003) Ablation of PLB exacerbates ischemic injury to a lesser extent in female than male mice: protective role of NO. Am J Physiol Heart Circ Physiol 284:H683–H690

Das B, Sarkar C (2006) Similarities between ischemic preconditioning and 17beta-estradiol mediated cardiomyocyte KATP channel activation leading to cardioprotective and antiarrhythmic effects during ischemia/reperfusion in the intact rabbit heart. J Cardiovasc Pharmacol 47:277–286

Dash R, Schmidt AG, Pathak A et al. (2003) Differential regulation of p38 mitogen-activated protein kinase mediates gender-dependent catecholamine-induced hypertrophy. Cardiovasc Res 57:704–714

Forster C, Kietz S, Hultenby K et al. (2004) Characterization of the ERbeta-/- mouse heart. Proc Natl Acad Sci USA 101:14234–14239

Gabel SA, Walker VR, London RE et al. (2005) Estrogen receptor beta mediates gender differences in ischemia/reperfusion injury. J Mol Cell Cardiol 38:289–297

Hsieh YC, Choudhry MA, Yu HP et al. (2006) Inhibition of cardiac PGC-1alpha expression abolishes ERbeta agonist-mediated cardioprotection following trauma-hemorrhage. FASEB J 20:1109–1117

Jazbutyte V, Hu K, Kruchten P et al. (2006) Aging reduces the efficacy of estrogen substitution to attenuate cardiac hypertrophy in female spontaneously hypertensive rats. Hypertension 48:579–586

Jovanovic S, Jovanovic A, Shen WK et al. (2000) Low concentrations of 17beta-estradiol protect single cardiac cells against metabolic stress-induced Ca^{2+} loading. J Am Coll Cardiol 36:948–952

Kadokami T, McTiernan CF, Kubota T et al. (2000) Sex-related survival differences in murine cardiomyopathy are associated with differences in TNF-receptor expression. J Clin Invest 106:589–597

Kadokami T, McTiernan CF, Hiquichi Y et al. (2005) 17 Beta-estradiol improves survival in male mice with cardiomyopathy induced by cardiac-specific tumor necrosis factor-alpha overexpression. J Interferon Cytokine Res 25:254–260

Kam KW, Qi JS, Chen M et al. (2004) Estrogen reduces cardiac injury and expression of beta1-adrenoceptor upon ischemic insult in the rat heart. J Pharmacol Exp Ther 309:8–15

Li Y, Kloner RA (1995) Is there a gender difference in infarct size and arrhythmias following experimental coronary occlusion and reperfusion? J Thromb Thrombolysis 2:221–225

Lobaton CD, Vay L, Hernandez-Sanmiguel E et al. (2005) Modulation of mitochondrial $Ca(2+)$ uptake by estrogen receptor agonists and antagonists. Br J Pharmacol 145:862–871

Murphy E, Perlman M, London RE et al. (1991) Amiloride delays the ischemia-induced rise in cytosolic free calcium. Circ Res 68:1250–1258

Nuedling S, Karas RH, Mendelshon ME et al. (2001) Activation of estrogen receptor beta is a prerequisite for estrogen-dependent upregulation of nitric oxide synthases in neonatal rat cardiac myocytes. FEBS Lett 502:103–108

Olsson MC, Palmer BM, Leinwand LA et al. (2001) Gender and aging in a transgenic mouse model of hypertrophic cardiomyopathy. Am J Physiol Heart Circ Physiol 280:H1136–H1144

Parkash J, Felty Q, Roy S (2006) Estrogen exerts a spatial and temporal influence on reactive oxygen species generation that precedes calcium uptake in high-capacity mitochondria: implications for rapid nongenomic signaling of cell growth. Biochemistry 45:2872–2881

Pedram A, Razandi M, Wallace DC et al. (2006) Functional estrogen receptors in the mitochondria of breast cancer cells. Mol Biol Cell 17:2125–2137

Pelzer T, Loza PA, Hu K et al. (2005) Increased mortality and aggravation of heart failure in estrogen receptor-beta knockout mice after myocardial infarction. Circulation 111:1492–1498

Peter I, Shearman AM, Vasan RS et al. (2006) Association of estrogen receptor beta gene polymorphisms with left ventricular mass and wall thickness in women. Am J Hypertens 18:1388–1395

Przyklenk K, Ovize M, Bauer B et al. (1995) Gender does not influence acute myocardial infarction in adult dogs. Am Heart J 129:1108–1113

Rossouw JE, Andersson GL, Prentice RL et al. (2002) Risks and benefits of estrogen plus progestin in healthy postmenopausal women: principal results From the Women's Health Initiative randomized controlled trial. JAMA 288:321–333

Schwend T, Gustafsson JA (2006) False positives in MALDI-TOF detection of ERbeta in mitochondria. Biochem Biophys Res Commun 343:707–711

Simoncini T, Hafezi-Moghadam A, Brazil DP et al. (2000) Interaction of oestrogen receptor with the regulatory subunit of phosphatidylinositol-3-OH kinase. Nature 407:538–541

Skavdahl M, Steenbergen C, Clark J et al. (2005) Estrogen receptor-beta mediates male-female differences in the development of pressure overload hypertrophy. Am J Physiol Heart Circ Physiol 288:H469–H476

Steenbergen C, Murphy E, Levy L et al. (1987) Elevation in cytosolic free calcium concentration early in myocardial ischemia in perfused rat heart. Circ Res 60:700–707

Sugishita K, Su Z, Li F et al. (2001) Gender influences [Ca(2+)](i) during metabolic inhibition in myocytes overexpressing the Na(+)-Ca(2+) exchanger. Circulation 104:2101–2106

Sun J, Picht E, Ginsburg KS et al. (2006) Hypercontractile female hearts exhibit increased S-nitrosylation of the L-type Ca2+ channel alpha1 subunit and reduced ischemia/reperfusion injury. Circ Res 98:403–411

Van Eickels M, Grohe C, Cleutjens JP et al. (2001) 17beta-estradiol attenuates the development of pressure-overload hypertrophy. Circulation 104:1419–1423

Wang M, Crisostomo P, Wariuko GM et al. (2006) Estrogen receptor-alpha mediates acute myocardial protection in females. Am J Physiol Heart Circ Physiol 290:H2204–H2209

Yang SH, Liu R, Perez EJ et al. (2004) Mitochondrial localization of estrogen receptor beta. Proc Natl Acad Sci USA 101:4130–4135

Yu HP, Shimizu T, Choudhry MA et al. (2006) Mechanism of cardioprotection following trauma-hemorrhagic shock by a selective estrogen receptor-beta agonist: up-regulation of cardiac heat shock factor-1 and heat shock proteins. J Mol Cell Cardiol 40:185–194

Zhai P, Eurell TE, Cooke PS et al. (2000) Myocardial ischemia-reperfusion injury in estrogen receptor-alpha knockout and wild-type mice. Am J Physiol Heart Circ Physiol 278:H1640–H1647

Ernst Schering Foundation Symposium Proceedings, Vol. 1, pp. 25–44
DOI 10.1007/2789_2006_015
© Springer-Verlag Berlin Heidelberg
Published Online: 4 May 2007

Genetic Dissection of Estrogen Receptor Signaling In Vivo

T.M. Wintermantel(✉), J. Elzer, A.E. Herbison, K.-H. Fritzemeier, G. Schütz

Therapeutic Research Group Gynecology and Andrology, Female Health Care Research, Bayer Schering Pharma AG, Müllerstr. 178, 13342 Berlin, Germany
email: tim.wintermantel@schering.de

1	Introduction	26
1.1	Estrogen Receptor α: What Have We Learned from Knock-Out Mice	26
1.2	Which Receptor for Which Estrogen Actions? ERα and ERβ, Novel ERs	27
1.3	Knock-Out Mice Revealed a Central Role of ERα in the Reproductive System	28
1.4	Challenges to Understanding Knock-Out Animal Models: Endocrine Regulation, Cell-Type-Specific Effects	30
2	The Cre-loxP System and Tissue-Specific Mutagenesis	30
3	Establishing a Conditional Allele for the Estrogen Receptor	32
3.1	Hepatocyte-Specific ERα Ablation	33
4	Ablation of the Estrogen Receptor in the Nervous System and Endocrine Regulation	34
4.1	Phenotype	34
4.2	Disrupted Estrogen-Positive Feedback in Mice Lacking ERα in Neurons	36
5	Knock-Out Mice and Isoform-Selective Agonists	37
5.1	In Vivo Observations Using ER-Isoform-Selective Steroids	38
5.2	ERα- and ERβ-Selective Ligands in Feedback	39
6	Summary	40
References		40

Abstract. The multiple actions of estrogen in mammalian physiology are brought about, on a molecular level, by several signaling pathways, and mediated by at least two receptors-estrogen receptor (ER) α and β. Analysis of knock-out mice devoid of either or both receptor isoforms revealed the essential function of estrogen receptor α in female reproduction, as ERα deficiency leads to a complex endocrine phenotype, severe disturbances in several reproductive organs, and infertility. This reflects the many actions of estrogen in female reproductive endocrinology. To carry the understanding of estrogen action to a cellular resolution, modern genetic technologies can be employed, including artificial chromosome-based transgenesis and conditional gene targeting. The combination of these techniques yields mouse models that lack ERα in specific cell types of the body. Using cell-type-specific ERα mutants, it could be shown that ERα in neurons is essential for the luteinizing hormone (LH) surge that triggers ovulation. Studies using ERα and ERβ-selective agonists reveal that ERα activation is sufficient to induce an ovulatory hormonal stimulus. Thus, genetic analysis and selective pharmacological tools can complement each other in the molecular and cellular dissection of hormone receptor function in vivo.

1 Introduction

1.1 Estrogen Receptor α: What Have We Learned from Knock-Out Mice

Apart from the hallmark functions of estradiol in maintaining female reproductive capacity-stimulation of uterine growth, mammary gland development and growth, and feedback regulation of the gonadotropin axis to establish the estrous cycle-it influences a number of processes in the body that are not connected to female reproduction. Its trophic actions on bone mass are well described, and its ability to prevent bone loss in postmenopausal women has been exploited clinically. While estradiol has vasculoprotective effects in preclinical models, recent clinical data have added controversy over whether this translates to clinical reality (Mendelsohn and Karas 2001; Turgeon et al. 2004). In cardiovascular disorders such as atherosclerosis, matters are complicated by a multimodal regulation of immune functions by estradiol (Carlsten 2005), and its influence on the blood lipid and protein profile (Blum and Cannon 1998).

In the central nervous system, the actions of estradiol exceed the endocrine feedback regulation, as estradiol has been shown to protect against brain injury, neurodegeneration, and cognitive decline. Clinical data suggest that estradiol reduces the risk of the onset and delays the progression of neurodegenerative diseases (van Amelsvoort et al. 2001).

The understanding of estrogen action both in the female reproductive tract and in other tissues has greatly profited from the analysis of mice with targeted mutations in the genes for the estradiol-producing enzyme aromatase (Jones et al. 2000; Simpson et al. 2005 and accompanying article by McPherson et al. and Simpson et al.), or in the two estrogen receptor isoforms estrogen receptor α (ERα) and estrogen receptor β (ERβ) (Couse and Korach 1999; Hewitt et al. 2004). In fact, a crucial role of the estrogen-ERα pathway in the male reproductive tract could be defined studying mice devoid of ERα (Eddy et al. 1996).

Nevertheless, challenges in estrogen research remain wherever the biological integration of estrogen actions involves hormone and receptor action in several organs (as in endocrine feedback regulation) or in different cell types and systems over a long time (e.g., in the cardiovascular system). In this article, we describe how novel genetic tools, such as cell-type-specific receptor ablation can be used to further the understanding of estrogen action in complex systems, and how genetic analysis can be complemented pharmacologically using receptor isoform-specific ligands.

1.2 Which Receptor for Which Estrogen Actions? ERα and ERβ, Novel ERs

The two known nuclear estrogen receptors in vertebrates, ERα and ERβ, are encoded by separate genes (*Esr1* and *Esr2*). Apart from the classical pathway (ligand diffusion, binding, transcriptional regulation) of nuclear hormone receptor action on hormone response elements (Beato et al. 1995), these receptors can also affect gene expression by influencing other transcription factors (Göttlicher et al. 1998), or regulate cytoplasmic signaling pathways upon binding to their ligand (so-called nongenomic or membrane-initiated steroid signaling (Hall et al. 2001)). A number of rapid estrogen actions could be demonstrated to rely on classical ERs, (see below). Prior to assigning a specific estrogen receptor signaling pathway to a biological action of estrogen, it is of course instrumental to identify the estrogen receptor isoform responsible for the biological effect under study. The aspect of assigning a biological action to a receptor isoform may sound trivial where one isoform is predominant (e.g., ERα in uterus, or ERβ in ovarian granulosa cells), but gains importance in tissues were both receptor isoforms are expressed (e.g., in brain) or where several cell types with varying expression of ERs interact (e.g., in the vessel wall). Furthermore, several alternative molecules have been suggested to mediate rapid actions of estrogens (Revankar et al. 2005; Valverde et al. 1999). The analysis of knock-out mice devoid of a specific isoform – and, once available, knock-out mice devoid of these alternative target structures – remains the gold standard in this crucial step during research. For instance, Abraham et al. (2004) used ER knock-out mice to show that estrogen-stimulated CREB phosphorylation in CNS neurons – a rapid, nongenomic effect – is mediated by classical ERs. This genetic analysis can be complemented, where possible, with receptor isoform-specific ligands (Bologa et al. 2006; Harris et al. 2003; Hillisch et al. 2004; Katzenellenbogen et al. 2001).

1.3 Knock-Out Mice Revealed a Central Role of ERα in the Reproductive System

The first available ERα knock-out mouse model (called ERαKO^{neo}) was generated by Korach and co-workers in 1993 (Lubahn et al. 1993). Lately, analysis of this "Chapel Hill" ERαKO line revealed an incomplete deletion of the ERα gene, as some tissues express a truncated form of the ERα protein (including the ligand-binding domain, yet devoid of the N-terminal activation function-1 (AF-1) (Pendaries et al. 2002) at low levels. The first ERβ knock-out mouse was established in 1998 (Krege et al. 1998). In 2000, newly targeted ERαKO and ERβKO lines were generated that showed a complete loss of the respective receptor (Dupont et al. 2000). The gene structure of steroid hormones can explain these differences in residual allele function: the first targeted mutations of steroid receptor alleles focussed on exon 2 (harboring the ATG start codon and AF1), but in some cases, transcripts can be generated including all exons from 3 onward. These transcripts retain both the DNA-binding domain (DBD) and the ligand-binding domain (LBD) (Cole et al. 1995; Mittelstadt and Ashwell 2003). Targeting exon 3, however, deletes the coding sequence for the very DNA-binding amino acids and thus abrogates DBD function. This strategy has been shown to result in nonfunctional alleles, as potential splice variants omitting the DBD should not have an open reading frame (Tronche et al. 1999; Wintermantel et al. 2006; T.M. Wintermantel, unpublished data).

This difference between the two ERαKO lines has shown to be of relevance in experiments addressing cardiovascular function (Pendaries et al. 2002), but their phenotypes are very similar in the female and male reproductive system, and very distinct from the ERβKO. Thus, in the remainder of this article, the term "ERαKO" will be used for either of the two lines.

Both sexes of the ERαKO are infertile, whereas only females of the ERβKO show a subfertile phenotype. The female ERαKO has immature ductal rudiments in the mammary tissue, whereas ERβKO females show a normal postnatal development of the mammary epithelial ducts (Forster et al. 2002; Mueller et al. 2002). ERαKO show elevated levels of LH, estrogen, and testosterone and low prolactin levels, while ERβKO mice have normal hormone levels (Couse et al. 2003). In the ERαKO ovary, hemorrhagic cystic follicles begin to develop at the beginning of puberty as a result of chronically elevated LH. In the ERβKO ovary, no signs of ovarian hyperstimulation are seen, but rather an impaired granulosa cell development (Couse et al. 2005; Krege et al. 1998). The uteri of ERαKO are severely hypoplastic and insensitive to estrogens, whereas ERβKO uteri show normal responses to estradiol treatment (Hewitt and Korach 2003). In summary, analysis of knock-out animals has revealed several essential functions for ERα in female fertility (HPG axis, uterus, mammary gland), whereas ERβ was shown to contribute to follicular granulosa cell function and may modulate HPG axis feedback (Dorling et al. 2003; for review, see Couse and Korach 1999; Hewitt et al. 2004). Studies in the cardiovascular system suggest specific, distinct roles for both receptors: ERα mediates estrogen effects on neointima formation in models of vessel injury (Pare et al. 2002), whereas ERβ seems to play a role in smooth muscle cells and blood pressure regulation (Zhu et al. 2002).

1.4 Challenges to Understanding Knock-Out Animal Models: Endocrine Regulation, Cell-Type-Specific Effects

The ERαKO model is a powerful tool for the understanding of ERα actions in mice, yet its contributions to the understanding of particular endocrine circuits and cell-type-specific events regulated by ERα is limited. In this model, the receptor is ablated in all tissues throughout development, so phenotypes can occur that represent the integration of several dysregulated circuits over time instead of.

Even though the ERαKO mouse phenotype shows that ERα is fundamental for the regulation of the hypothalamic-pituitary-gonadal (HPG) axis, it remains to be understood how and on which cell type of the HPG axis ERα acts. In the hypothalamus, GnRH neurons themselves are ERα-negative (Herbison and Pape 2001). In the pituitary, LH but not follicle-stimulating hormone (FSH) secretion is influenced by ERα (Couse et al. 2003). It is the combined ablation of estrogen signaling in the hypothalamus and the pituitary that leads to the anovulatory ERαKO phenotype, with the ovulation process itself not being dependent on ERα action locally (Couse et al. 1999).

It has been demonstrated that estrogen has neuroprotective effects in models of brain ischemia, and that these effects are lost in the ERαKO (Dubal et al. 2001). But it remains to be determined whether this neuroprotective effect is mediated by ERα action in neurons, microglial cells, or endothelial cells.

In a different context, it is unclear whether ERα prevents atherosclerosis by regulating cells of the immune system or if it acts primarily on endothelial or smooth muscle cells of the vascular system (Mendelsohn and Karas 1999).

These examples show the necessity of tissue- and cell-type-selective techniques to investigate ERα action on a cellular level in vivo.

2 The Cre-loxP System and Tissue-Specific Mutagenesis

To address the issues mentioned above, the Cre-loxP recombination system can be used. It allows the ablation of the gene of interest in any desired cell population without affecting the gene's function in other organs.

The Cre-loxP-system was first described in bacteriophages and consists of two components: a sequence-specific recombinase (Cre, a 36-kDa protein) and a DNA sequence flanked by loxP sites (34-bp DNA elements, which are recognized the Cre-recombinase) (Nagy 2000). The Cre-recombinase catalyzes recombination of two loxP sites. There are no other compounds or cofactors necessary to catalyze this reaction, which has been demonstrated to work in *Escherichia coli*, yeast, plants, and animal organisms. Flanking an essential exon of a gene

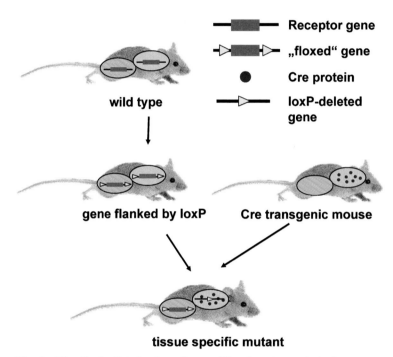

Fig. 1. The Cre-loxP technology for conditional mutagenesis in the mouse. The gene of interest is floxed by gene-targeting techniques. This mouse line is bred to a transgenic mouse expressing Cre under the control of a tissue-specific promotor. This leads to gene deletion only in those cells or tissues that express Cre recombinase. (Reproduced with permission from Wintermantel et al. (2004) by Georg Thieme)

by loxP sites and recombination of these sites by the Cre-recombinase leads to loss of the exon sequence and thereby to the loss of the cell's capability to translate a functional protein. In vivo, gene targeting enables the loxP tagging of a gene of interest. Cell-type-specific expression of the Cre-recombinase is achieved by a transgenic construct to drive transcription of a Cre-cDNA under control of a tissue-specific promotor. Breeding of a Cre-transgenic mouse line to a second mouse line harboring the loxP-tagged gene of interest results in a mouse model that lacks the gene of interest in a distinct cell population only (Fig. 1).

This technology allows the investigation of the role of a gene in a certain organ or cell type without distortion caused by systemic influences, as well as the role of a hormone receptor in a specific organ in a complex endocrine circuit.

3 Establishing a Conditional Allele for the Estrogen Receptor

To investigate tissue-specific actions of the ERα using cell-type-specific gene ablation, a conditional allele of the *Esr1* allele in the mouse was generated by gene targeting (Wintermantel et al. 2006): the *Esr1* gene in mouse embryonic stem cells was modified to carry loxP sites surrounding exon 3 (Fig. 2a; see above). From these embryonic stem cells, the ERαfl mouse line was established. The functionality of this allele was shown first by normal ERα expression and fertility in mice homozygous for the ERαfl allele, then by loss of ERα protein after Cre-mediated recombination (Fig. 2; see below). Finally, mice homozygous for a germline deletion of the ERαfl gene (achieved by breeding the ERαfl line to a Cre deleter) show a phenotype very similar to the phenotype described for ERαKO knock-out animals, including polycystic ovaries and uterine hypoplasia (T.M. Wintermantel and G. Schütz, unpublished data).

Using the ERαfl mouse line (or the similar line described by Dupont et al. 2000) in conjunction with a Cre-expressing transgenic mouse line, the function of ERα in different cell types can be studied in the context of an intact organism with normal reproductive physiology, including normal estrogen levels and estrous cycle.

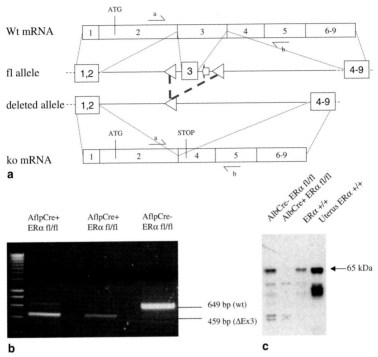

Fig. 2a–c. Generation and analysis of a conditional allele for mouse ERα: **a** The floxed allele (*second line*) gives rise to the wild-type mRNA (*top line*). After Cre-mediated recombination and excision of exon 3, the deleted allele (*third line*) is unable to generate more than a knock-out mRNA (*fourth line*) lacking an open reading frame. **b,c** Hepatocyte-specific ablation of ERα. RT-PCR (**b**) with primers depicted in **a** shows normal signal from wt mRNA in fl/fl controls, and a faint band that stems from the deleted allele. **c** Liver of ERαfl/fl; AlfpCre+ mice (*second lane*) is devoid of ERα protein, as judged by Western blot

3.1 Hepatocyte-Specific ERα Ablation

By breeding the ERαfl mouse line to the AlfpCre mutant mouse line (Kellendonk et al. 2000), ablation of the ERα could be achieved in hepatocytes (Fig. 2b,c; T.M. Wintermantel and G. Schütz, unpublished data). These mice are fertile and show no overt phenotype. In contrast to the ERα knock-out mice, which are obese (Ohlsson et al. 2000), liver-specific ERα knock-out mice do not develop obesity, nor do they show any histopathological alteration in livers of either male or female mutants. Whereas the reaction of liver-specific ERα knock-out mice to metabolic stimuli or stress remains to be investigated, this mouse line demonstrates that, under basal conditions, hepatic estrogen receptor α does not play a critical role in body weight regulation, or, in other words, that the lack of ERα action in liver does not specifically contribute to the obese phenotype of the ERαKO mouse.

4 Ablation of the Estrogen Receptor in the Nervous System and Endocrine Regulation

Mice carrying a deletion of ERα in neuronal cells were generated by breeding the ERαfl line to the CaMKIIαCre-transgenic mouse line (Casanova et al. 2001). These mice lack ERα specifically in neurons of the central nervous system, whereas the receptor protein is still present in other cells, including the anterior pituitary (see Fig. 3a–d; Wintermantel et al. 2006).

4.1 Phenotype

Adult female brain-specific ERα-mutant mice were found to be infertile and exhibit striking abnormalities in their reproductive organs (Fig. 3e,f): in the mutants, the uterus was grossly enlarged and filled with liquid, with the endometrium proper being severely atrophic and

Fig. 3a–h. Generation and phenotype of neuron-specific ERα mutants. **a,c** ERα immunoreactivity in hypothalamus and pituitary of control ERαfl/fl mice. **b** Lack of detectable ERα protein in the hypothalamus of neuron-specific ERαfl/fl; CaMKCre[+] mutant mice, yet normal ERα expression in the pituitary (**d**). **f,h** Characteristic uterine and ovarian phenotype (see text) in neuron-specific ERαfl/fl; CaMKCre[+] mutant mice, as compared to ERαfl/fl controls (**e,g**). **e,f** Mouse uterus upon dissection. *Inset* H&E staining of endometrium shows lack of glandular structures in mutants (**f**). **g,h** H&E stainings of ovaries. (Adapted from Wintermantel et al. 2006)

Genetic Dissection of Estrogen Receptor Signaling In Vivo 35

ERα fl/fl   ERα fl/fl CamKCre

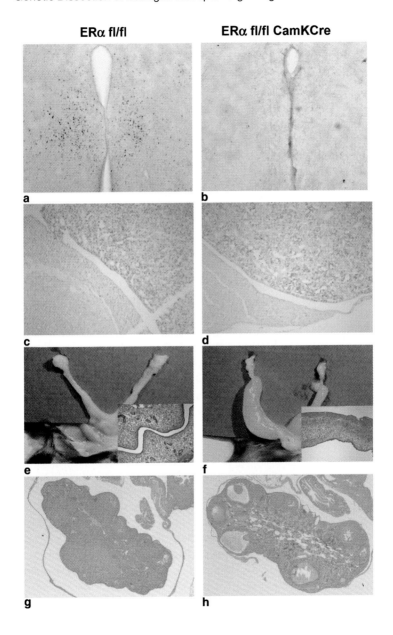

devoid of glandular structures, with signs of granulocyte infiltration. This phenotype is first observed at the age of 5 weeks, suggesting a connection with the onset of ovarian steroid synthesis. In fact, after ovariectomy, the uterine inflammation regresses (data not shown). In line with this hypothesis, ovaries of brain-specific mutant mice showed histological signs of gonadotropin hyperactivation: in mutant mice, ovaries contained a large number of antral follicles compared with wild-type animals (Fig. 3g,h). Interestingly, we did not observe corpora lutea in the mutants, suggesting an inability of the mutants to ovulate.

4.2 Disrupted Estrogen-Positive Feedback in Mice Lacking ERα in Neurons

In order to establish whether estrogen feedback leading to ovulation was still intact in these mice, estrogen-positive gonadotropin regulation was investigated.

Whereas wild-type, homozygous ERαfl controls or CaMKIIαCre transgenic mice harboring wild-type ERα alleles were able to mount an LH surge following estrogen stimulation several days after ovariectomy and estrogen replacement, the neuron-specific ERα knock-out animals failed to exhibit the LH surge. This correlated with an inability to activate GnRH neurons in the preoptic hypothalamus, as assessed by c-fos staining. Control mice, in contrast, show robust activation of GnRH neurons after estrogen stimulation, concomitant with the LH surge. Similarly, ERβ knock-out mice are fully capable of activating GnRH neurons and eliciting an LH surge when examined in the same experimental setting (Wintermantel et al. 2006).

These results demonstrate that ERα in neurons is essential for estrogen-positive feedback and, thus, for physiological ovulation to occur. The infertile reproductive phenotype of neuron-specific ERα mutant mice can entirely be explained by the absence of estrogen-positive feedback. Ovarian histology shows an absence of corpora lutea and abundance of antral follicles supporting the failure of the ovulatory mechanism. Unlike in the global ERαKO (Dupont et al. 2000; Lubahn et al. 1993), the uteri of neuron-specific ERα mutant mice remain sensitive to estrogen and can be ectopically stimulated by tonically (noncycling) elevated estrogen levels. Thus, the reproductive phenotype of the neuron-specific ERα mutant mouse model is compatible with that of an animal lacking positive feedback. This would enable relatively normal basal gonadotrophin secretion from the pituitary, sufficient to promote follicular growth and estrogen production. In contrast to the global ERαKO, estrogen-negative feedback on the pituitary is still possible in this model, restraining circulating LH and preventing a hemorrhagic follicular phenotype (Wintermantel et al. 2006). Yet in the absence of ERα signaling in the brain, no positive feedback, no gonadotropin surge, and no ovulation is possible.

5   Knock-Out Mice and Isoform-Selective Agonists

Genetic loss-of-function models have been powerful tools for the identification of biological processes regulated by a certain gene, or, as shown in the case of the ERα/ERβ mutant, to identify the gene or receptor (out of more than one isoform) responsible for the biological process under study. One caveat, however, remains for the interpretation even of highly cell-type-specific gene ablations: the phenotype of a mutant model is subject to developmental processes that could, in turn, depend on the very gene that is ablated. The role of ERα, for example, could be to contribute to the establishment of the neuronal network responsible for positive feedback – and ERβ action, in turn, could be important for feedback in a wild-type context where it has not been lacking throughout development. To answer these questions, selective receptor ligands (chemical genetics) have been an important tool. It should be kept in mind that, on the one hand, specificity and in vivo selectivity of chemical compounds is always a source of uncertainty – for many specific kinase inhibitors, off-target effects were identified (Daub et al. 2004). This fact and the identification of novel receptors or target structures for classical hormones (with different pharmacological properties) might cause reinterpretation of data obtained in experiments where a synthetic compound was used to study a specific gene's or receptor's biological action. However, when combining pharmacological and genetic data, the specific strengths of both approaches can be seized: the unambiguousness of a genetic loss-of-function experiment, and the greater versatility (concerning time control and species usage) of pharmacological experiments.

5.1 In Vivo Observations Using ER-Isoform-Selective Steroids

The steroidal estrogen receptor-subtype-selective agonists 16α-LE$_2$ (ERα-selective) and 8β-VE$_2$ (ERβ-selective) have been designed using molecular modeling structures (Fig. 4). When their in vivo pharmacological properties were studied, interesting insights into the function of the two estrogen receptor isoforms were obtained in several experimental models in the rat (Hegele-Hartung et al. 2004; Hillisch et al. 2004):

In line with the high expression of ERβ in granulosa cells of the ovary (Kuiper et al. 1996) and the functional impairment of the ERβ knockout ovarian follicle (Couse et al. 2005; Emmen et al. 2005), the ERβ-selective agonist 8β-VE$_2$ has been found to stimulate ovarian follicle development in hypophysectomized rats (Hegele-Hartung et al. 2004). The ERα-selective agonist, 16α-LE$_2$-stimulated uterine weight, where 8β-VE$_2$ was ineffective at isoform-selective doses (Hillisch et al. 2004).

16αL-E$_2$

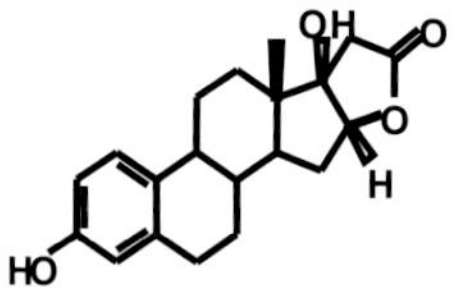

8βV-E$_2$

300fold ERα-selective	190fold ERβ-selective
stimulates uterine growth	stimulates ovarian follicles
activates GnRH neurons	no activation of GnRH neurons
triggers LH surge	no LH surge

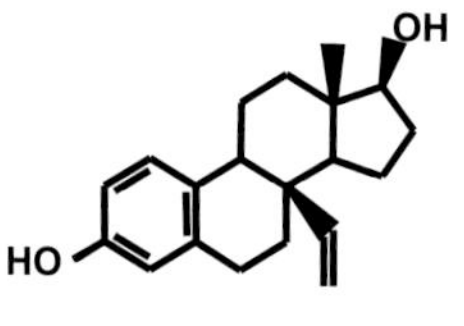

Fig. 4. Structures of 3,17-Dihydroxy-19-nor-17 alpha-pregna-1,3,5(10)-triene-21,16 alpha-lactone (16αL-E$_2$, ERα-selective, *left*) and 8-Vinylestra-1,3,5(10)-triene-3,17 beta-diol (8βV-E$_2$, ERβ-selective, *right*), along with a summary of the specific actions discussed in the text (Hillisch et al. 2004)

These data are in good agreement with the studies in ERα and ERβ knock-out mice, where it was shown that the uterotrophic responses to estradiol are primarily mediated via ERα, and occur normally in the absence of ERβ (Hewitt and Korach 2003). Similarly, 16α-LE$_2$ reduced serum cholesterol (Hillisch et al. 2004), in line with studies using ERα knock-out mice that showed an essential function of this receptor in cholesterol homeostasis (Ohlsson et al. 2000).

5.2 ERα- and ERβ-Selective Ligands in Feedback

The studies in global ERα, ERβ, and neuron-specific ERα knock-out mice (see Sect. 4 above) indicated that the presence of ERα in neurons is *necessary* for positive feedback. To investigate whether the signaling of ERα is *sufficient*, the steroidal subtype-selective agonists 16α-LE$_2$ and 8β-VE$_2$ were studied in wildtype mice under the same experimental paradigm for positive feedback used for the knock-out mice. These experiments showed that 16α-LE$_2$ was capable of eliciting both an LH surge and the activation of GnRH neurons, whereas 8β-VE$_2$ was not (Wintermantel et al. 2006; T.M. Wintermantel, R. Porteous and A.E. Herbison, unpublished data), which not only is consistent with the data obtained from knock-out mice, but also establishes that ERα signaling in neurons is both necessary and sufficient for estrogen-positive feedback to occur.

In the global ERα knock-out mice, basal LH levels are highly elevated, leading to hemorrhages and cysts in the ovary. This effect is not seen in the neuron-specific ERα knock-out mice, indicating that ERα signaling in the pituitary, which is still intact in these mice yet absent in the global ERα knock-out mice, is capable of controlling basal LH levels (whereas no LH peak can occur in either of the two models). A role for ERα in pituitary estrogen-negative gonadotropin feedback is also suggested by experiments using ovariectomized rats: in this model, 16α-LE$_2$ suppresses the increase in LH and FSH evoked by ovariectomy, whereas 8β-VE$_2$ is not effective at selective doses (Hillisch et al. 2004). In summary, the concurrent genetic and pharmacological data highlight the specific role of ERα in endocrine regulation.

6 Summary

The molecular understanding of estrogen action has greatly profited from the analysis of ERαKO and ERβKO mice. To investigate complex aspects of endocrine physiology in closer detail and cellular resolution, tissue-specific mutagenesis is an important tool. The analysis of more- and more detailed-cell-type specific estrogen receptor mutant models will undoubtedly add valuable information about estrogen signaling in complex physiological and pathological processes, e.g., atherosclerosis or estrogen-mediated vasculo- and neuroprotection. The availability of subtype-selective ER ligands can add important insight to these studies. After the definition of a receptor isoform responsible for estrogen action in a given process, the next crucial step in estrogen receptor research will be to dissect the different signaling pathways of the receptor. This will involve both mechanism-selective mutations, and, ultimately, mechanism-selective ligands.

References

Abraham IM, Todman MG, Korach KS, Herbison AE (2004) Critical in vivo roles for classical estrogen receptors in rapid estrogen actions on intracellular signaling in mouse brain. Endocrinology 145:3055–3061

Beato M, Herrlich P, Schütz G (1995) Steroid hormone receptors: many actors in search of a plot. Cell 83:851–857

Blum A, Cannon ROI (1998) Effects of oestrogens and selective oestrogen receptor modulators on serum lipoproteins and vascular function. Curr Opin Lipidol 9:575–586

Bologa CG, Revankar CM, Young SM, Edwards BS, Arterburn JB, Kiselyov AS, Parker MA, Tkachenko SE, Savchuck NP, Sklar LA et al. (2006) Virtual and biomolecular screening converge on a selective agonist for GPRN30. Nat Chem Biol 2:207–212

Carlsten H (2005) Immune responses and bone loss: the estrogen connection. Immunol Rev 208:194–206

Casanova E, Fehsenfeld S, Mantamadiotis T, Lemberger T, Greiner E, Stewart AF, Schutz G (2001) A CamKIIalpha iCre BAC allows brain-specific gene inactivation. Genesis 31:37–42

Cole TJ, Blendy JA, Monaghan AP, Krieglstein K, Schmid W, Aguzzi A, Fantuzzi G, Hummler E, Unsicker K, Schutz G (1995) Targeted disruption of the glucocorticoid receptor gene blocks adrenergic chromaffin cell development and severely retards lung maturation. Genes Dev 9:1608–1621

Couse JF, Korach KS (1999) Estrogen receptor null mice: what have we learned and where will they lead us? Endocr Rev 20:358–417

Couse JF, Bunch DO, Lindzey J, Schomberg DW, Korach KS (1999) Prevention of the polycystic ovarian phenotype and characterization of ovulatory capacity in the estrogen receptor-α knockout mouse. Endocrinology 140:5855–5865

Couse JF, Yates MM, Walker VR, Korach KS (2003) Characterization of the hypothalamic-pituitary-gonadal axis in estrogen receptor (ER) null mice reveals hypergonadism and endocrine sex reversal in females lacking ERalpha but not ERbeta. Mol Endocrinol 17:1039–1053

Couse JF, Yates MM, Deroo BJ, Korach KS (2005) Estrogen receptor-{beta} is critical to granulosa cell differentiation and the ovulatory response to gonadotropins. Endocrinology 146:3244–3246

Daub H, Specht K, Ullrich A (2004) Strategies to overcome resistance to targeted protein kinase inhibitors. Nat Rev Drug Discov 3:1001–1010

Dorling AA, Todman MG, Korach KS, Herbison AE (2003) Critical role for estrogen receptor alpha in negative feedback regulation of gonadotropin-releasing hormone mRNA expression in the female mouse. Neuroendocrinology 78:204–209

Dubal DB, Zhu H, Yu J, Rau SW, Shughrue PJ, Merchenthaler I, Kindy MS, Wise PM (2001) Estrogen receptor alpha, not beta, is a critical link in estradiol-mediated protection against brain injury. Proc Natl Acad Sci USA 98:1952–1957

Dupont S, Krust A, Gansmuller A, Dierich A, Chambon P, Mark M (2000) Effect of single and compound knockouts of estrogen receptors α (ERα) and β (ERβ) on mouse reproductive phenotypes. Development 127:4277–4291

Eddy EM, Washburn TF, Bunch DO, Goulding EH, Gladen BC, Lubahn DB, Korach KS (1996) Targeted disruption of the estrogen receptor gene in male mice causes alteration of spermatogenesis and infertility. Endocrinology 137:4796–4805

Emmen JM, Couse JF, Elmore SA, Yates MM, Kissling GE, Korach KS (2005) In vitro growth and ovulation of follicles from ovaries of estrogen receptor (ER)-{alpha} and ER{beta} null mice indicate a role for ER{beta} in follicular maturation. Endocrinology 146:2817–2826

Forster C, Makela S, Warri A, Kietz S, Becker D, Hultenby K, Warner M, Gustafsson JA (2002) Involvement of estrogen receptor beta in terminal differentiation of mammary gland epithelium. Proc Natl Acad Sci USA 99:15578–15583

Göttlicher M, Heck S, Herrlich P (1998) Transcriptional cross-talk, the second mode of steroid hormone receptor action. J Mol Med 76:480–489

Hall JM, Couse JF, Korach KS (2001) The multifaceted mechanisms of estradiol and estrogen receptor signaling. J Biol Chem 276:36869–36872

Harris HA, Albert LM, Leathurby Y, Malamas MS, Mewshaw RE, Miller CP, Kharode YP, Marzolf J, Komm BS, Winneker RC et al. (2003) Evaluation of an estrogen receptor-beta agonist in animal models of human disease. Endocrinology 144:4241–4249

Hegele-Hartung C, Siebel P, Peters O, Kosemund D, Muller G, Hillisch A, Walter A, Kraetzschmar J, Fritzemeier KH (2004) Impact of isotype-selective estrogen receptor agonists on ovarian function. Proc Natl Acad Sci USA 101:5129–5134

Herbison AE, Pape JR (2001) New evidence for estrogen receptors in gonadotropin-releasing hormone neurons. Front Neuroendocrinol 22:292–308

Hewitt SC, Korach KS (2003) Oestrogen receptor knockout mice: roles for oestrogen receptors alpha and beta in reproductive tissues. Reproduction 125:143–149

Hewitt SC, Harrell JC, Korach KS (2004) Lessons in estrogen biology from knockout and transgenic animals. Annu Rev Physiol 67:285–308

Hillisch A, Peters O, Kosemund D, Muller G, Walter A, Schneider B, Reddersen G, Elger W, Fritzemeier KH (2004) Dissecting physiological roles of estrogen receptor alpha and beta with potent selective ligands from structure-based design. Mol Endocrinol 18:1599–1609

Jones ME, Thorburn AW, Britt KL, Hewitt KN, Wreford NG, Proietto J, Oz OK, Leury BJ, Robertson KM, Yao S, Simpson ER (2000) Aromatase-deficient (ArKO) mice have a phenotype of increased adiposity. Proc Natl Acad Sci USA 97:12735–12740

Katzenellenbogen BS, Sun J, Harrington WR, Kraichely DM, Ganessunker D, Katzenellenbogen JA (2001) Structure-function relationships in estrogen receptors and the characterization of novel selective estrogen receptor modulators with unique pharmacological profiles. Ann NY Acad Sci 949:6–15

Kellendonk C, Opherk C, Anlag K, Schütz G, Tronche F (2000) Hepatocyte-specific expression of Cre recombinase. Genesis 26:151–153

Krege JH, Hodgin JB, Couse JF, Enmark E, Warner M, Mahler JF, Sar M, Korach KS, Gustafsson J-A, Smithies O (1998) Generation and reproductive phenotypes of mice lacking estrogen receptor β. Proc Natl Acad Sci USA 95:15677–15682

Kuiper GG JM, Enmark E, Pelto-Huikko M, Nilsson S, Gustafsson J-A (1996) Cloning of a novel estrogen receptor expressed in rat prostate and ovary. Proc Natl Acad Sci USA 93:5925–5930

Lubahn D, Moyer JS, Golding TS, Couse JF, Korach KS, Smithies O (1993) Alteration of reproductive function but not prenatal sexual development after insertional disruption of the mouse estrogen receptor gene. Proc Natl Acad Sci USA 90:11162–11166

Mendelsohn ME, Karas RH (1999) The protective effects of estrogen on the cardiovascular system. N Engl J Med 340:1801–1811

Mendelsohn ME, Karas RH (2001) The time has come to stop letting the HERS tale wag the dogma. Circulation 104:2256–2259

Mittelstadt PR, Ashwell JD (2003) Disruption of glucocorticoid receptor exon 2 yields a ligand-responsive C-terminal fragment that regulates gene expression. Mol Endocrinol 17:1534–1542

Mueller SO, Clark JA, Myers PH, Korach KS (2002) Mammary gland development in adult mice requires epithelial and stromal estrogen receptor alpha. Endocrinology 143:2357–2365

Nagy A (2000) Cre recombinase: the universal reagent for genome tailoring. Genesis 26:99–109

Ohlsson C, Hellberg N, Parini P, Vidal O, Bohlooly M, Rudling M, Lindberg MK, Warner M, Angelin B, Gustafsson J-A (2000) Obesity and disturbed lipoprotein profile in estrogen receptor-alpha-deficient male mice. Biochem Biophys Res Commun 278:640–645

Pare G, Krust A, Karas RH, Dupont S, Aronovitz M, Chambon P, Mendelsohn ME (2002) Estrogen receptor-alpha mediates the protective effects of estrogen against vascular injury. Circ Res 90:1087–1092

Pendaries C, Darblade B, Rochaix P, Krust A, Chambon P, Korach KS, Bayard F, Arnal JF (2002) The AF-1 activation-function of ERalpha may be dispensable to mediate the effect of estradiol on endothelial NO production in mice. Proc Natl Acad Sci USA 99:2205–2210

Revankar CM, Cimino DF, Sklar LA, Arterburn JB, Prossnitz ER (2005) A transmembrane intracellular estrogen receptor mediates rapid cell signaling. Science 307:1625–1630

Simpson ER, Misso M, Hewitt KN, Hill RA, Boon WC, Jones ME, Kovacic A, Zhou J, Clyne CD (2005) Estrogen – the good, the bad, and the unexpected. Endocr Rev 26:322–330

Tronche F, Kellendonk C, Kretz O, Gass P, Anlag K, Orban PC, Bock R, Klein R, Schutz G (1999) Disruption of the glucocorticoid receptor gene in the nervous system results in reduced anxiety. Nat Genet 23:99–103

Turgeon JL, McDonnell DP, Martin KA, Wise PM (2004) Hormone therapy: physiological complexity belies therapeutic simplicity. Science 304:1269–1273

Valverde MA, Rojas P, Amigo J, Cosmelli D, Orio P, Bahamonde MI, Mann GE, Vergara C, Latorre R (1999) Acute activation of maxi-K channels (hSlo) by estradiol binding to the β subunit. Science 285:1929–1931

Van Amelsvoort T, Compton J, Murphy D (2001) In vivo assessment of the effects of estrogen on human brain. Trends Endocrinol Metab 12:273–276

Wintermantel TM, Berger S, Greiner EF, Schutz G (2004) Genetic dissection of corticosteroid receptor function in mice. Hormone Metab Res 36:387–391

Wintermantel TM, Campbell RE, Porteous R, Bock D, Gröne H-J, Todman MG, Korach KS, Greiner E, Perez CA, Schuetz G, Herbison AE (2006) Definition of estrogen receptor pathway critical for estrogen positive feedback to gonadotropin-releasing hormone neurons and fertility. Neuron 52:271–280

Zhu Y, Bian Z, Lu P, Karas RH, Bao L, Cox D, Hodgin J, Shaul PW, Thoren P, Smithies O et al. (2002) Abnormal vascular function and hypertension in mice deficient in estrogen receptor beta. Science 295:505–508

Of Mice and Men: The Many Guises of Estrogens

E.R. Simpson(✉), M.E. Jones

Prince Henry's Institute of Medical Research, P.O. Box 5152, VIC 3168 Clayton, Australia

email: *evan.simpson@princehenrys.org*

1	The Concept of Local Estrogen Biosynthesis	46
2	Aromatase and Its Gene	50
3	Aromatase Deficiency	50
4	Consequences of Estrogen Deficiency in Males	53
5	Estrogen in Skeletal Growth, Maturation, and Maintenance	53
6	Energy Homeostasis and the Metabolic Syndrome	55
6.1	Adiposity	55
6.2	Insulin Resistance	56
7	Estrogens and Male Fertility	57
8	Brain and Behavior Phenotypes	58
9	Cardiovascular System	59
10	Implications for Use of Aromatase Inhibitors	60
11	Conclusions	61
References		62

Abstract. Models of estrogen insufficiency have revealed new and unexpected roles for estrogens in males as well as females. These models include natural mutations in the aromatase gene in humans, as well as mouse knock-outs of aromatase and the estrogen receptors, and one man with a mutation in the ERα gene. These mutations, both natural and experimental, have revealed that estrogen deficiency results in a spectrum of symptoms. These include loss of fertility and libido in both males and females; loss of bone in both males and females; a cardiovascular and cerebrovascular phenotype; development of a metabolic syndrome in both males and females, with truncal adiposity and male-specific hepatic steatosis. Most of these symptoms can be reversed or attenuated by estradiol therapy. Thus estrogen is involved in the maintenance of general physiological homeostasis in both sexes.

1 The Concept of Local Estrogen Biosynthesis

Models of estrogen insufficiency have revealed new and unexpected roles for estrogens in both males and females. These models include natural mutations in the aromatase gene in humans, as well as mouse knock-outs of aromatase and the estrogen receptors (Lubahn et al. 1993; Krege et al. 1998; Couse et al. 1999; Dupont et al. 2000; Fisher et al. 1998; Jones et al. 2000). In addition, there is one man described with a natural mutation in ERα (Smith et al. 1994). Some of the roles of estrogens apply to both males and females and do not relate to reproduction, for example the bone, vascular, and metabolic syndrome phenotypes.

In postmenopausal women, in whom the ovaries cease to produce estrogens, and in men, estradiol does not appear to function to any great extent as a circulating hormone; instead, it is synthesized in a number of extragonadal sites such as breast, brain, and bone, where its actions are mainly at the local level as a paracrine or intracrine factor. Thus, in postmenopausal women and in men, circulating estrogens are not the drivers of estrogen action; instead, they reflect the metabolism of estrogens formed in these extragonadal sites; they are reactive rather than proactive (Labrie et al. 2003). Importantly, estrogen biosynthesis in these sites depends on a circulating source of androgenic precursors such as testosterone.

Table 1 shows the plasma steroid levels in postmenopausal women and in men. As can be seen, the levels of estrone and estradiol in the plasma of postmenopausal women are extremely low, lower in fact than those in the circulation of men; and moreover, the levels of circulating testosterone are an order of magnitude greater than those of estrogens in postmenopausal women. This in itself would suggest that circulating testosterone is better placed to serve as a precursor of functional estra-

Table 1 Plasma steroid levels in postmenopausal women and in men

	Women	(nmol/L)	Men
T	0.6		12
Δ^4	2.5		4
E1	0.10		0.13
E2	0.04		0.10
DHEA	15		10
DHEAS	2500		2000

diol in target tissues than is circulating estradiol. On the other hand, the levels of testosterone in the blood of men are an order of magnitude higher than those of women. Significantly, levels of dehydroepiandrosterone (DHEA) and DHEA sulfate (DHEAS) in the blood of both men and women are orders of magnitude higher than those of the circulating active steroids. In postmenopausal women, the ovaries secrete 25%–35% of the circulating testosterone. The remainder is formed peripherally from androstenedione and DHEA produced in the ovaries and from androstenedione, DHEA, and DHEAS secreted by the adrenals. However, the secretion of these steroids and their plasma concentrations decrease markedly with advancing age (Labrie et al. 1997).

Figure 1 shows the metabolism of testosterone and estradiol in a typical target cell (Labrie et al. 2003). Testosterone in this cell can be derived from the uptake of testosterone or of androstenedione, DHEA, or DHEAS, all of which may be converted in the target cell to testosterone. Testosterone in turn can act directly on the androgen receptor or be converted to dihydrotestosterone, which then acts on the androgen receptor. Alternatively, testosterone can be converted to estradiol, which in turn acts on the estrogen receptor. Testosterone and estradiol can then leave the cell as such or be converted to reduced and conjugated metabolites that circulate in the blood at concentrations higher than those of the active steroids (Labrie et al. 2003). Based on these considerations, it is difficult to see how one can readily equate plasma levels of testosterone and estradiol to the concentrations that are present in target cells. These considerations lead to the following conclusions regarding the significance of peripheral steroid metabolism:

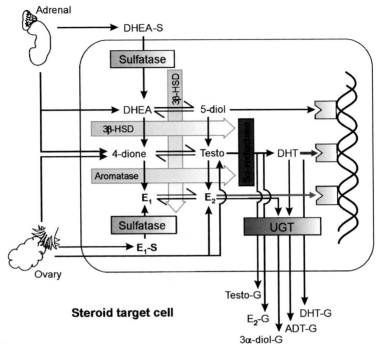

Fig. 1. Pathways of metabolism of testosterone and estradiol in target tissues. *HSD*, Hydroxysteroid dehydrogenase; *5-diol*, 5/-androstanediol; *4-dione*, androstenedione; *testo*, testosterone; *E1*, estrone; *E2*, estradiol; *DHT*, dihydrotestosterone; *UGT*, UDP-glucuronyl transferase; *G*, glucuronate. (Reproduced with permission from Labrie et al. 2003)

1. Women and men make closer to equal amounts of testosterone and estradiol than previously thought (on the order of 30%–40% rather than 10% in the case of women relative to men), and both hormones have major physiological roles in both sexes.
2. However, in premenopausal women, most of the testosterone is formed, acts, and is metabolized in specific target tissues. Testosterone is a paracrine and intracrine factor, whereas in men it circulates as a hormone and acts globally.

3. On the other hand, in men most of the estradiol is formed, acts, and is metabolized in specific target tissues, whereas in women it circulates as a hormone and acts globally.
4. Finally in postmenopausal women, neither testosterone nor estradiol functions to any extent as a circulating hormone. Both are mainly formed locally in target tissues and act and are metabolized therein.

The power of local estrogen biosynthesis is illustrated in the case of postmenopausal breast cancer (Pasqualini et al. 1996). It has been determined that the concentration of estradiol present in breast tumors of postmenopausal women is at least 20-fold greater than that present in the plasma. With aromatase inhibitor therapy, there is a precipitous drop in the intratumoral concentrations of estradiol and estrone together with a corresponding loss of intratumoral aromatase activity, consistent with this activity within the tumor and the surrounding breast adipose tissue being responsible for these high tissue concentrations (DeJong et al. 1997).

In bone, aromatase is expressed primarily in osteoblasts and chondrocytes (Oz et al. 2001), and aromatase activity in cultured osteoblasts is comparable to that present in adipose stromal cells (Shozu and Simpson 1998). Thus, it appears that in bone also, local aromatase expression is a major source of estrogen responsible for the maintenance of mineralization, although this is extremely difficult to prove because of sampling problems. Hence, for both breast tumors and for bone, it is likely that circulating estrogen levels are minimally responsible for the relatively high endogenous tissue estrogen levels; rather, the circulating levels reflect the sum of local formation in its various sites. This is a fundamental concept for the interpretation of relationships between circulating estrogen levels in postmenopausal women and estrogen insufficiency or excess in specific tissues.

The second important point is that estrogen production in these extragonadal sites is dependent on an external source of C19 androgenic precursors, because these extragonadal tissues are incapable of converting cholesterol to the C19 steroids (Labrie et al. 1997, 1998). As a consequence, circulating levels of testosterone and androstenedione as well as DHEA and DHEAS become extremely important in terms of providing adequate substrate for estrogen biosynthesis in these sites, and therefore differences in the levels of circulating androgens are likely to be important determinants for the maintenance of local estrogen levels in extragonadal sites. In this context, it is appropriate to consider why osteoporosis is more common in women than in men and why it affects women at a younger age in terms of fracture incidence. We have suggested that uninterrupted sufficiency of circulating testosterone in men throughout life supports the local production of estradiol by aromatization of testosterone in estrogen-dependent tissues, and thus affords ongoing protection against the so-called estrogen deficiency diseases. This appears to be important in terms of protecting the bones of men against mineral loss and may also contribute to the maintenance of cognitive function and prevention of Alzheimer's disease (Simpson et al. 2000).

2 Aromatase and Its Gene

Estrogen biosynthesis is catalyzed by a microsomal member of the cytochrome P450 superfamily, namely aromatase cytochrome P450 (P450arom, the product of the *CYP19* gene (Means et al. 1989; Harada et al. 1990; Toda et al.1990; Simpson et al. 2002). The P450 gene superfamily is a very large one, containing over 3000 members in some 350 families, of which cytochrome P450arom is the sole member of family 19 (website of D. Nelson, http://drnelson.utmem.edu/cytochromeP450.html). This heme protein is responsible for binding of the C19 androgenic steroid substrate and catalyzing the series of reactions leading to formation of the phenolic A ring characteristic of estrogens.

3 Aromatase Deficiency

Remarkably few clinical cases have been reported of humans in whom, as a result of a natural mutation, aromatase is nonfunctional. To date, only seven males (Maffei et al. 2004; Morishima et al. 1995; Carani et al. 1997; Hermann et al. 2002; Bouillon et al. 2004; Deladoey et al. 1999; Pura et al. 2003) and six females (Simpson 2004) have been reported with aromatase deficiency. Affected females present at birth with ambiguous genitalia, and again at puberty with primary amenorrhea, failure of breast development, hypergonadotropic hypogonadism, and cystic ovaries. Symptoms regress with estrogen treatment but, as a result of this therapy, the long-term sequelae of aromatase deficiency have not been studied in women. Consequently this review focuses on the male phenotype.

All mutations accounting for the aromatase deficiency in affected males have been located within exons V or IX in the CYP19 gene (Fig. 2). These regions are unlikely to represent mutational hotspots in the aromatase gene but rather regions encoding essential functions in the aromatization process. Indeed, exon V encodes residues essential for the catalytic activity of the enzyme, whereas the highly conserved exon IX contains the substrate-binding domain (Chen and Zhou 1992).

Each of these different mutations generates an inactive aromatase enzyme, and therefore, not surprisingly, all adult male patients demonstrate common phenotypic features that include undetectable estrogens; normal to high levels of testosterone and gonadotropins; tall stature with delayed skeletal maturation and epiphyseal closure, and eunuchoid body proportions; and osteoporosis with bone pain and progressive genu

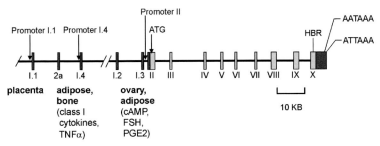

Fig. 2. Diagram of the human aromatase (*CYP19*) gene showing tissue-specific promoter usage. The coding region comprises exons II–X. Upstream of the translational start site (*ATG*) are a number of untranslated exons I that are spliced into the coding region at a common three-splice junction in a tissue-specific fashion due to use of the promoters I.1–I.4. The promoters are regulated by the factors indicated. Because this splice junction is upstream of the start of translation, the coding region is always the same, regardless of the tissue of expression. *HBR*, heme-binding region

valgum (knock knees). In addition, there are reports of hyperinsulinemia, impaired lipid metabolism, and impaired reproductive function (Grumbach and Auchus 1999; Faustini-Fustini et al. 1999; Rochira et al. 2002) (Tables 2, 3). However, in order to identify and understand the biochemical and molecular pathways orchestrating the actions of es-

Table 2 Metabolic and Liver Function Parameters of the Argentinian Patient

	Before E treatment	After E treatment
Total cholesterol (mg/dl)	177	110
LDL cholesterol	107	66
HDL cholesterol	31	41
Triglycerides	199	106
Glucose (70–110 mg/dl)	180	144
Insulin (5–30 µU/ml)	94	53
Fructosamine (µmol/l)	406	315
Liver function parameters:		
GPT (< 37 U/l)	195	70
GOT (< 40 U/l)	108	45
γ-GT (< 11–50 U/l)	153	42

E, Estrogen; HDL, high-density lipoprotein; LDL, low-density lipoprotein; GPT, glutamate Pyruvate transaminase; GOT, glutamate oxaloacetate transaminase; γ-GT, γ-glutamyltransferase

Table 3 ArKO mice develop insulin resistance

	Insulin (m/Ul)	Glucose (mmol/l)
ArKO		
4 months old	5.98 ± 1.00 (3)	N.D.
1 year old	38.67 ± 11.18 (5)*	8.52 ± 1.56 (3)
WT		
4 months old	5.26 ± 0.75 (4)	N.D.
1 year old	13.82 ± 3.82 (4)	8.61 ± 2.02 (3)

Data represent mean $\pm$ SEM (n) N.D., Not determined

trogen in males, it is necessary to look to animal models such as the aromatase-(ArKO) and estrogen receptor (ER$^-$)-knock-out mice.

4 Consequences of Estrogen Deficiency in Males

Estrogen deficiency has an impact on many body systems. Comparing estrogen-deficient men and mice will enable a deeper understanding of the roles of estrogen in males. The common phenotypic features of human males have been discussed above. Mice homozygous for the disrupted aromatase gene (i.e., ArKO mice) are born phenotypically normal. In males, testosterone levels were variable but generally elevated, as were serum levels of 5α-dihydrotestosterone (Fisher et al. 1998; McPherson et al. 2001; Robertson et al. 1999). Luteinizing hormone (LH) levels were significantly elevated (Fisher et al. 1998), but follicle stimulating hormone (FSH) levels were unchanged (Robertson et al. 1999).

5 Estrogen in Skeletal Growth, Maturation, and Maintenance

The importance of estrogen in maintaining bone mass in females is firmly established, but estrogens also have an essential role in male bone homeostasis, as indicated by studies in aromatase-deficient men and mice. As indicated above, aromatase-deficient men are extremely tall (>3 standard deviations; Oettel 2002), with continued linear growth into adulthood, unfused epiphyses, increased bone remodeling, and reduced bone mineral density (BMD) (Oettel 2002; Gennari et al. 2004; Khosla et al. 2002). Estradiol (E$_2$) treatment caused cessation of linear growth concomitantly with epiphyseal fusion, reduced bone turnover and increased bone density (Carani et al. 1997; Bouillon et al. 2004; Rochira et al. 2002; Khosla et al. 2002; Herrmann et al. 2005). Up until the 1990s, it was believed that testosterone regulated bone metabolism and pubertal bone growth in men, and estrogen was not considered a player in the process (Khosla et al. 2002). However, this data generated from the aromatase-deficient men contributed significantly to the growing body of evidence demonstrating the integral role of estrogen in bone growth and homeostasis in men (Gennari et al. 2004). Indeed, the local conversion of circulating testosterone to active estrogen by bone-specific aromatase is positively correlated with bone maintenance (Simpson et al. 2002). The most recently diagnosed aromatase-deficient male provided valuable information on the relative contribution of estrogen to bone growth around the time of puberty (Bouillon et al. 2004). Daily administration of E_2 between the ages of 17 years, 1 month, and 20 years, 5 months, increased lumbar spine and femoral neck BMD by 23% and 14, respectively, and the bone age increased from 12 years, 5 months to 16 years, 8 months. Body height increased by 8.5% and bone size increased substantially. Continued linear growth with delayed epiphyseal closure probably reflected the younger bone age at which E_2 treatment was initiated, and the authors suggest that this provides evidence that estrogens accelerate skeletal growth during puberty, and then produce a growth-limiting effect through epiphyseal closure at cessation of the pubertal transition.

A single case of estrogen insensitivity has been reported in a man with a mutated ERα (Smith et al. 1994). Despite elevated levels of circulating estrogen in the presence of normal testosterone, the estrogen was ineffective. As with the aromatase-deficient men, this man was tall and suffered from skeletal dysfunction – unfused epiphyses, osteoporosis, delayed bone age, increased bone turnover, and progressive genu valgum – further evidence that a healthy male skeleton requires active estrogen and that the actions of E_2 are likely to be mediated through ERα.

Male ArKO mice at 5–7 months of age showed reduced femur length compared with age-matched wild-type littermates (Oz et al. 2000), with an absence of the expected femur growth acceleration during puberty (Oz et al. 2001). This is the opposite effect to that occurring in men with aromatase deficiency, who are tall as a result of the failure of their epiphyseal plates to close. Reduced bone mass was evident in ArKO male lumbar vertebrae, with significant decreases in trabecular bone volume and thickness, concomitant with a low bone turnover characterized by decreases in osteoblastic, osteoclastic, and mineralizing surfaces compared with wild-type males, and supported by significant decreases in serum osteocalcin levels (Oz et al. 2000). By contrast, a second group reported increased bone turnover in male ArKO mice from their colony

(Miyaura et al. 2001), which is in accordance with that observed in aromatase-deficient men. E_2 restored bone mass in these mice (Miyaura et al. 2001). An explanation of the disparity in bone turnover indices reported for these two colonies of ArKO mice is currently unclear, but it is still possible to conclude that estrogen "profoundly influences processes associated with bone turnover" (Oz et al. 2000; Miyaura et al. 2001). Similarly to ArKO males, ERα-knock-out- (αERKO) and double ER-knock-out (DERKO) mice have shortened femur length, and αERKO males exhibit increased bone turnover (Khosla et al. 2002). Skeletons of βERKO males appear normal, further confirming that ERα is the primary receptor mediating the actions of E_2 on bone (Khosla et al. 2002). This is consistent with the presence of skeletal and bone turnover defects in the male human with a selective mutation in ERα who had high endogenous estrogen levels and a normally functioning ERβ.

6 Energy Homeostasis and the Metabolic Syndrome

6.1 Adiposity

Aromatase deficiency in men is associated with a perturbed lipid profile (Faustini-Fustini et al. 1999) (Table 2), and patients present with body mass indexes (BMI) primarily in the overweight range (BMI 25–30 kg/m^2) (Maffei et al. 2004; Morishima et al. 1995; Carani et al. 1997; Hermann et al. 2002; Pura et al. 2003) with an accumulation of abdominal adipose tissue. Circulating levels of triglycerides generally are elevated, with low circulating high-density lipoprotein (HDL) cholesterol. Overall, lipid-associated parameters improved following E_2 treatment; however, results were variable, probably as a consequence of varying treatment regimens or variations in genetic background. Indeed, the patient from Argentina (Maffei et al. 2004) received alendronate and supraphysiological doses of testosterone sequentially for 37 months before diagnosis of aromatase deficiency. Hence, baseline variables before E_2 treatment might not be a true reflection of an aromatase-deficient profile. With this in mind, after a no-treatment phase of 9 months before beginning E_2 therapy, total cholesterol and triglyceride levels were normal, and HDL cholesterol was decreased (Table 2).

Severe hepatic steatosis has been observed in male, but not female, ArKO mice (Jones et al. 2000; Takeda et al. 2003; Hewitt et al. 2003) as a consequence of elevated hepatic triglyceride (sevenfold) and cholesterol (~40%) levels (Hewitt et al. 2003). The retention of lipids in the liver was accompanied by an increase in the expression of enzymes involved in de novo lipogenesis, including fatty acid synthase and acetyl coenzyme A carboxylase α, and the fatty acid transporter, adipocyte differentiation-related protein (Hewitt et al. 2003). E_2 administration tended to reduce the expression of these genes and the hepatic steatosis was ameliorated (Hewitt et al. 2004). Other researchers reported impaired fatty acid β-oxidation enzyme activity in lipid-laden male ArKO livers; this was similarly improved following administration of E_2 (Toda et al. 2001b). Hepatic steatosis has been reported in two aromatase-deficient men (Maffei et al. 2004; Pura et al. 2003), associated with an enlarged liver (Pura et al. 2003) and with significantly elevated liver enzymes (Maffei et al. 2004; Pura et al. 2003) [glutamic pyruvic transaminase, 195 U/l; glutamic oxaloacetic transaminase, 108 U/l; and γ-glutamyl transferase, 153 U/l (Maffei et al. 2004)]. Following 12 months of E_2 treatment in one of these patients, each of these parameters had either improved or recovered to within normal parameters (Maffei et al. 2004). Together, these data present compelling evidence that estrogen has a pivotal role in maintaining lipid homeostasis in males at the level of gene expression as well as cellular sequelae.

6.2 Insulin Resistance

The relationship between estrogen and glucose metabolism and insulin action was first indicated in the estrogen-resistant ERα-mutated male who presented with insulin resistance (Smith et al. 1994). This relationship has been confirmed in the subsequent analyses of aromatase-deficient men (Fisher et al. 1998; Freedman et al. 2005; Gennari et al. 2004) (Table 1), with only one exception (Gennari et al. 2004). An elevated homeostasis model assessment of insulin resistance (HOMA IR), a well-accepted parameter for insulin resistance determination, calculated as [fasting insulin (μU/ml) × fasting plasma glucose (mmol/l) × 22.5] value of 3.6 (normal < 2.4), indicated modest insulin resistance in one of these men (Grumbach and Auchus 1999), and another was diagnosed with type 2 diabetes mellitus with elevated fasting glucose and insulin levels (180 mg/dl and 94 μU/ml, respectively) (Maffei et al. 2004). E_2 treatment improved the insulin sensitivity in both of these men (Maffei et al. 2004; Hermann et al. 2002).

ArKO mice are insulin-resistant, with elevated circulating levels of insulin concomitant with normoglycemia in 1-year-old ArKO males compared with wild-type animals (Jones et al. 2000) (Table 3). This has been confirmed by Takeda et al. (2003), who demonstrated that ArKO male mice developed glucose intolerance and insulin resistance in an age-dependent manner, beginning at 18 weeks of age, which was ameliorated by E_2 administration. Treatment with the cholesterol-lowering fibrate bezafibrate or the insulin-sensitizing thiazolidinedione pioglitazone also improved insulin resistance, although neither compound restored sensitivity to wild-type levels (Takeda et al. 2003).

Insulin resistance is characteristic of obesity, and there exists a complex relationship between estrogen, adiposity, and glucose metabolism. Clearly, estrogen has a significant role in maintaining energy homeostasis in males, including glucose tolerance and insulin action.

7 Estrogens and Male Fertility

Although there is general uniformity between most characteristics of aromatase-deficient men, there are no consistent findings with respect to the individual parameters of the testes of these same individuals (Table 2). Semen analyses have been performed on three patients (Carani et al. 1997; Hermann et al. 2002; Pura et al. 2003), two patients having been excluded from analyses owing to prepubertal age (Deladoey et al. 1999) and for religious reasons (Maffei et al. 2004). One patient refused testing (Morishima et al. 1995). Hence, primarily as a result of so few cases of aromatase deficiency in men, no firm conclusions can be drawn regarding the exact nature of the role of estrogen in the human testis. However, impairment of fertility was common to four of the adult male patients with congenital estrogen deficiency, suggesting a strong relationship between estrogen deficiency and compromised fertility in males. With respect to the ER-mutated individual, sperm density was normal but with decreased viability (Smith et al. 1994).

Without exception, all of the aromatase-deficient men investigated were heterosexual, and libido and sexual functioning, assessed by interview and self-reported diary, were reported as being normal (Maffei et al. 2004; Morishima et al. 1995; Carani et al. 1997; Hermann et al. 2002), suggesting that endogenous estrogens are not essential for male sexual behavior. However, only one was married and none had children. Moreover, reassessment of two of these men following estrogen therapy demonstrated a significant improvement in libido and frequency of sexual activity (Carani et al. 1999, 2005), suggesting that estrogen might act synergistically with androgen to enhance sexual behavior in men.

Characterization of the male reproductive system of the ArKO mice has been comprehensively reviewed by O'Donnell et al. (2001). Briefly, the male mice exhibit an age-progressive infertility, and by 1 year of age the epididymides of some animals were devoid of sperm (Fisher et al. 1998; Toda et al. 2001a). When spermatozoa were present in the epididymides of 1-year-old ArKO mice, there was a significant decrease in sperm concentration and motility, and sperm were unable to fertilize oocytes in vitro (Robertson et al. 2001), although Toda et al. (2001a) reported that at 10 months of age, the sperm from ArKO mice generated in their laboratory were able to fertilize oocytes *in* vitro. All three lobes of the ArKO prostate – ventral, anterior and dorsolateral – weighed significantly more than those of age-matched wild-type littermates from 8 to 56 weeks of age. This is a consequence of hyperplasia of the epithelial, interstitial, and luminal compartments, rather than aberrant growth or malignancy (McPherson et al. 2001; Toda et al. 2001a).

8 Brain and Behavior Phenotypes

Aromatase (Lephart et al. 2001; Wagner and Morrell 1997) and ERs (Mitra et al. 2003) are known to be expressed in numerous regions of the brain, such as the hypothalamus, amygdala, and hippocampus, suggesting an important role for estrogen in the brain. By 1 year of age, in the male ArKO mouse, tyrosine hydroxylase-immunopositive neurons (tyrosine hydroxylase being a marker for dopaminergic neurons) in the medial preoptic area (MPO) and arcuate nucleus (Arc) were undergoing apoptosis, whereas no such observations were made in wild-type littermates. Concomitantly, the number of tyrosine hydroxylase-positive neurons in the MPO and Arc were observed to be significantly lowered in these ArKO animals (Hill et al. 2004). Thus, these data showed that estrogen is essential to the integrity and survival of dopaminergic neurons in the MPO and Arc of male mice. This could be mediated via upregulation of the expression of bcl-2-related anti-apoptotic genes (Pile 1999; Dong et al. 1999) by estrogens. Indeed, RNase protection assays indicated that the ratio of bcl-2-related anti-apoptotic gene transcripts to pro-apoptotic gene transcripts was significantly downregulated in ArKO mice.

Because the dopamine released in the MPO was reported to be an important neurotransmitter for male copulatory behavior (Hull et al. 1995), the decrease in dopaminergic neurons in the ArKO mouse is likely to have an impact on its sexual behavior. Indeed, male ArKO mice display disruption in their sexual behavior (Robertson et al. 2001), which can be partially restored by E_2 treatment in adulthood (Bakker et al. 2004). The MPO and Arc have also been implicated in the regulation of aggressive behavior (Olivier et al. 1983), and therefore the lack of aggression in the ArKO mice (Robertson et al. 2001) might be a result of the destruction in these regions in the estrogen-deficient state.

9 Cardiovascular System

It has long been believed that estrogen provides cardioprotective benefits for women, and evidence is growing that estrogen is also required for normal cardiovascular functioning in males (Komesaroff and Sudhir 2001). Vascular system anomalies have been examined in only one aromatase-deficient patient (Maffei et al. 2004). The carotid arteries showed two lipid plaques 4 mm and 3 mm thick, which completely disappeared following 12 months of E_2 treatment (Maffei et al. 2004).

Preliminary examination has been made of the ArKO cardiovascular system. Utilizing isolated male ArKO aortic ring segments in isometric myographs, our collaborators have demonstrated that loss of estrogen results in a diminished response to the endothelium-dependent agonist acetylcholine, a response shown to be dependent on nitric oxide production (Kimura et al. 2003). These findings suggest that estrogen facilitates vasorelaxation in males. We have also examined the role of endogenous estrogens on the proliferation and apoptosis of vascular smooth muscle cells from ArKO males. These cells display diminution to growth stimuli and are more susceptible to apoptosis in vitro; both parameters being corrected by exogenous estrogen supplementation (Ling et al. 2004).

Given that cardiovascular disease is the primary cause of death in both men and women (American Heart Association 2006), and that men are at higher risk of developing cardiovascular disease than are premenopausal women (Reckelhoff 2001), the importance of this model in understanding the link between estrogen and cardiovascular health is apparent. These two studies are the beginning of such research.

10 Implications for Use of Aromatase Inhibitors

Clinical use of aromatase inhibitors (AIs) is increasing rapidly as third-generation AIs prove efficacious in treating and preventing breast cancer in postmenopausal women (Freedman et al. 2005; Osborne and Tripathy 2005). In men, AIs have been, or are being, examined as potential therapies for breast cancer (Zabolotny et al. 2005), gynecomastia (Plourde et al. 2004; Riepe et al. 2004; Boccardo et al. 2005; Rhoden and Morgentaler 2004), hypogonadism (Dougherty et al. 2005; Leder et al. 2004; Holbrook and Cohen 2003), short stature and bone metabolism in pubertal boys (Mauras et al. 2004; Zhou et al. 2005; Dunkel and Wickman 2003; Wickman et al. 2003), epilepsy (Harden and MacLusky 2004), infertility (Raman and Schlegel 2003), and androgen-independent prostate carcinoma (Smith et al. 2003; Santen et al. 2001). Yet little is known about the long-term side effects of AI treatment in either sex. It has been established in women that AI therapy results in loss of BMD and increased fracture risk, and that this can be alleviated with bisphosphonate therapy, which inhibits osteoclast-mediated bone resorption (Chlebowski 2005). MetS risk factors have not been reported as a consequence of AI treatment, but it is generally agreed that long-term data on side effects and safety are required before conclusions can be drawn (Osborne and Tripathy 2005; Howell and Cuzick 2005). Indeed, breast cancer therapy with tamoxifen, a nonsteroidal estrogen antagonist, induced hepatic steatosis as a complication (Gunel et al. 2003; Nishino et al. 2003; Nemoto et al. 2003) and was strongly associated with one or more of the MetS risk factors: insulin resistance, central obesity, and hypertriglyceridemia (Nemoto et al. 2003; Bruno et al. 2005). Interestingly, tamoxifen-induced hepatic steatosis has been demonstrated to be closely correlated with polymorphisms in the *CYP17* gene (Ohnishi et al. 2005). *CYP17*, similarly to *CYP19*, encodes an enzyme pivotal to estrogen synthesis, 17α-hydroxylase. Hence, before long-term data is available, it is reasonable to look to our current models of aromatase deficiency as indicators of potential side effects of extended AI use.

11 Conclusions

Significant aspects of the phenotypic features of the male ArKO mouse model closely align with the clinical profile of aromatase-deficient men. We have summarized the available data on the consequences of estrogen deficiency in men and mice, reporting that such diverse (but integrated) systems as the brain, bones, cardiovascular, immune, and reproductive systems and energy homeostasis all have impaired functions in the absence of estrogen. The information provided by the aromatase-deficient men and ArKO mice will advance our fundamental understanding of the role of estrogen in maintaining male health. Continued research will undoubtedly yield further novel (and perhaps unexpected) functions for estrogen in the male. A major goal for ongoing and future research is to resolve which of the two ER isoforms is involved in regulation of the various physiological systems whose function is altered as a consequence of estrogen deficiency. This issue is currently under intense study in several laboratories. The outcome is likely to be the therapeutic use of ER isoform-specific agonists (and possibly, antagonists) to provide specific interventions in estrogen-deficient conditions.

References

American Heart Association (2006) Heart Disease and Stroke Statistics: 2006 Update. American Heart Association, Dallas

Bakker J et al (2004) Restoration of male sexual behavior by adult exogenous estrogens in male aromatase knockout mice. Horm Behav 46:1–10

Boccardo F et al (2005) Evaluation of tamoxifen and anastrozole in the prevention of gynecomastia and breast pain induced by bicalutamide monotherapy of prostate cancer. J Clin Oncol 23:808–815

Bouillon R et al (2004) Estrogens are essential for male pubertal periosteal bone expansion. J Clin Endocrinol Metab 89:6025–6029

Bruno S, Maisonneuve P, Castellana P et al (2005) Incidence and risk factors for non-alcoholic steatohepatitis: prospective study of 5408 women enrolled in Italian tamoxifen chemoprevention trial. BMJ 330:932

Carani C et al (1997) Effect of testosterone and estradiol in a man with aromatase deficiency. N Engl J Med 337:91–95

Carani C et al (1999) Role of oestrogen in male sexual behaviour: insights from the natural model of aromatase deficiency. Clin Endocrinol (Oxf) 51:517–524

Carani C et al (2005) Sex steroids and sexual desire in a man with a novel mutation of aromatase gene and hypogonadism. Psychoneuroendocrinology 30:413–417

Chen S, Zhou D (1992) Functional domains of aromatase cytochrome P450 inferred from comparative analyses of amino acid sequences and substantiated by site-directed mutagenesis experiments. J Biol Chem 267:22587–22594

Chlebowski RT (2005) Bone health in women with early-stage breast cancer. Clin Breast Cancer 5(Suppl 2):S35–S40

Couse JF, Hewitt SC, Bunch DO et al (1999) Postnatal sex reversal of the ovaries in mice lacking estrogen receptors α and β. Science 286:2328–2331

DeJong PC, ven de Ven J, Nortier HW et al (1997) Inhibition of breast cancer tissue aromatase activity and estrogen concentrations by the third-generation aromatase inhibitor vorozole. Cancer Res 57:2109–2111

Deladoey J et al (1999) Aromatase deficiency caused by a novel P450arom gene mutation: impact of absent estrogen production on serum gonadotropin concentration in a boy. J Clin Endocrinol Metab 84:4050–4054

Dong L et al (1999) Mechanisms of transcriptional activation of bcl-2 gene expression by 17beta-estradiol in breast cancer cells. J Biol Chem 274:32099–32107

Dougherty RH et al (2005) Effect of aromatase inhibition on lipids and inflammatory markers of cardiovascular disease in elderly men with low testosterone levels. Clin Endocrinol (Oxf) 62:228–235

Dunkel L, Wickman S (2003) Novel treatment of short stature with aromatase inhibitors. J Steroid Biochem Mol Biol 86:345–356

Dupont S, Krust A, Gansmuller A et al (2000) Effect of single and compound knockouts of estrogen receptor α (ERα) and β (ERβ) on mouse reproductive phenotypes. Development 127:4277–4291

Faustini-Fustini M et al (1999) Oestrogen deficiency in men: where are we today? Eur J Endocrinol 140:111–129

Fisher CR, Graves KH, Parlow AF, Simpson ER (1998) Characterization of mice deficient in aromatase (ArKO) because of targeted disruption of the cyp19 gene. Proc Natl Acad Sci USA 95:6965–6970

Freedman OC et al (2005) Using aromatase inhibitors in the neoadjuvant setting: evolution or revolution? Cancer Treat Rev 31:1–17

Gennari L et al (2004) Aromatase activity and bone homeostasis in men. J Clin Endocrinol Metab 89:5898–5907

Grumbach MM, Auchus RJ (1999) Estrogen: consequences and implications of human mutations in synthesis and action. J Clin Endocrinol Metab 84:4677–4694

Gunel N et al (2003) Serum leptin levels are associated with tamoxifen-hepatic steatosis. Curr Med Res Opin 19:47–50

Harada N, Yamada K, Saito K et al (1990) Structural characterization of the human estrogen synthetase (aromatase gene). Biochem Biophys Res Commun 166:365–372

Harden C, MacLusky NJ (2004) Aromatase inhibition, testosterone and seizures. Epilepsy Behav 5:260–263

Hermann BL et al (2002) Impact of estrogen replacement therapy in a male with congenital aromatase deficiency caused by a novel mutation in the CYP19 gene. J Clin Endocrinol Metab 87:5476–5484

Herrmann BL et al (2005) Effects of estrogen replacement therapy on bone and glucose metabolism in a male with congenital aromatase deficiency. Horm Metab Res 37:178–183

Hewitt KN et al (2003) The aromatase knockout mouse presents with a sexually dimorphic disruption to cholesterol homeostasis. Endocrinology 144:3895–3903

Hewitt KN et al (2004) Estrogen replacement reverses the hepatic steatosis phenotype in the male aromatase knockout mouse. Endocrinology 145:1842–1848

Hill RA et al (2004) Estrogen deficiency leads to apoptosis in dopaminergic neurons in the medial preoptic area and arcuate nucleus of male mice. Mol Cell Neurosci 27:466–476

Holbrook JM, Cohen PG (2003) Aromatase inhibition for the treatment of idiopathic hypogonadotropic hypogonadism in men with premature ejaculation. South Med J 96:544–547

Howell A, Cuzick J (2005) Vascular effects of aromatase inhibitors: data from clinical trials. J Steroid Biochem Mol Biol 95:143–149

Hull EM et al (1995) Extracellular dopamine in the medial preoptic area: implications for sexual motivation and hormonal control of copulation. J Neurosci 15:7465–7471

Jones MEE, Thorburn AW, Britt KL et al (2000) Aromatase-deficient (ArKO) mice have a phenotype of increased adiposity. Proc Natl Acad Sci USA 97:12735–12740

Khosla S et al (2002) Clinical review 144: estrogen and the male skeleton. J Clin Endocrinol Metab 87:1443–1450

Kimura M et al (2003) Impaired acetylcholine-induced release of nitric oxide in the aorta of male aromatase-knockout mice: regulation of nitric oxide production by endogenous sex hormones in males. Circ Res 93:1267–1271

Komesaroff PA, Sudhir K (2001) Estrogens and human cardiovascular physiology. Reprod Fertil Dev 13:261–272

Krege JH, Hodgin JB, Couse JF et al (1998) Generation and reproductive phenotypes of mice lacking estrogen receptor-β. Proc Natl Acad Sci USA 95:15677–15682

Labrie F, Belanger A, Cusan L et al (1997) Marked decline in serum concentrations of adrenal C19 sex steroid precursor and conjugated androgen metabolites during aging. J Clin Endocrinol Metab 82:2396–2402

Labrie F, Belanger A, Luu-The V et al (1998) DHEA and the intracrine formation of androgens and estrogens in peripheral target tissues: its role during aging. Steroids 63:322–328

Labrie F, Luu-The V, Labrie C et al (2003) Endocrine and intracrine sources of androgens in women: inhibition of breast cancer and other roles of androgens and their precursor dehydroepiandrosterone. Endocr Rev 24:152–182

Leder BZ et al (2004) Effects of aromatase inhibition in elderly men with low or borderline-low serum testosterone levels. J Clin Endocrinol Metab 89:1174–1180

Lephart ED et al (2001) Brain androgen and progesterone metabolizing enzymes: biosynthesis, distribution and function. Brain Res Brain Res Rev 37:25–37

Ling S et al (2004) Endogenous estrogen deficiency reduces proliferation and enhances apoptosis-related death in vascular smooth muscle cells: insights from the aromatase-knockout mouse. Circulation 109:537–543

Lubahn DB, Moyer JS, Golding TS et al (1993) Alteration of reproductive function but not prenatal sexual development after insertional disruption of the mouse estroven receptor gene. Proc Natl Acad Sci USA 90:11162–11166

Maffei L et al (2004) Dysmetabolic syndrome in a man with a novel mutation of the aromatase gene: effects of testosterone, alendronate, and estradiol treatment. J Clin Endocrinol Metab 89:61–70

Mauras N et al (2004) An open label 12-month pilot trial on the effects of the aromatase inhibitor anastrozole in growth hormone (GH)-treated GH deficient adolescent boys. J Pediatr Endocrinol Metab 17:1597–1606

McPherson SJ et al (2001) Elevated androgens and prolactin in aromatase-deficient mice cause enlargement, but not malignancy, of the prostate gland. Endocrinology 142:2458–2467

Means GD, Mahendroo M, Corbin CJ et al (1989) Structural analysis of the gene encoding human aromatase cytochrome P-450, the enzyme responsible for estrogen biosynthesis. J Biol Chem 264:19385–19391

Mitra SW et al (2003) Immunolocalization of estrogen receptor beta in the mouse brain: comparison with estrogen receptor alpha. Endocrinology 144:2055–2067

Miyaura C et al (2001) Sex- and age-related response to aromatase deficiency in bone. Biochem Biophys Res Commun 280:1062–1068

Morishima A et al (1995) Aromatase deficiency in male and female siblings caused by a novel mutation and the physiological role of estrogens. J Clin Endocrinol Metab 80:3689–3698

Nemoto Y et al (2003) Tamoxifen-induced nonalcoholic steatohepatitis in breast cancer patients treated with adjuvant tamoxifen. Intern Med 41:345–350

Nishino M et al (2003) Effects of tamoxifen on hepatic fat content and the development of hepatic steatosis in patients with breast cancer: high frequency of involvement and rapid reversal after completion of tamoxifen therapy. AJR Am J Roentgenol 180:129–134

O'Donnell L et al (2001) Estrogen and spermatogenesis. Endocr Rev 22:289–318

Oettel M (2002) Is there a role for estrogens in the maintenance of men's health? Aging Male 5:248–257

Ohnishi T et al (2005) CYP17 polymorphism as a risk factor of tamoxifen-induced hepatic steatosis in breast cancer patients. Oncol Rep 13:485–489

Olivier B et al (1983) Effect of anterior hypothalamic and mammillary area lesions on territorial aggressive behaviour in male rats. Behav Brain Res 9:59–81

Osborne C, Tripathy D (2005) Aromatase inhibitors: rationale and use in breast cancer. Annu Rev Med 56:103–116

Oz OK et al (2000) Bone has a sexually dimorphic response to aromatase deficiency. J Bone Miner Res 15:507–514

Oz OK et al (2001) Bone phenotype of the aromatase deficient mouse. J Steroid Biochem Mol Biol 79:49–59

Oz OK, Millsaps R, Welch R et al (2001) Expression of aromatase in the human growth plate. J Mol Endocrinol 27:249–253

Pasqualini JR, Chetrite G, Blacker C et al (1996) Concentrations of estrone, estradiol, and estrone sulfate and evaluation of sulfatase and aromatase activities in pre- and postmenopausal breast cancer patients. J Clin Endocrinol Metab 81:1460–1464

Pile CJ (1999) Estrogen modulates neuronal Bcl-cL expression and beta-amyloid-induced apoptosis: relevance to Alzheimer's disease. J Neurochem 72:1552–1563

Plourde PV et al (2004) Safety and efficacy of anastrozole for the treatment of pubertal gynecomastia: a randomized, double-blind, placebo-controlled trial. J Clin Endocrinol Metab 89:4428–4433

Pura M et al (2003) Clinical findings in an adult man with a novel mutation in the aromatase gene. In 85th Annual Meeting The Endocrine Society, Philadelphia, PA

Raman JD, Schlegel PN (2003) Aromatase inhibitors for male infertility. J Urol 167:624–628

Reckelhoff JF (2001) Gender differences in the regulation of blood pressure. Hypertension 37:1199–1208

Rhoden EL, Morgentaler A (2004) Treatment of testosterone-induced gynecomastia with the aromatase inhibitor, anastrozole. Int J Impot Res 16:95–97

Riepe FE et al (2004) Treatment of pubertal gynecomastia with the specific aromatase inhibitor anastrozole. Horm Res 62:113–118

Robertson KM et al (1999) Impairment of spermatogenesis in mice lacking a functional aromatase (cyp19) gene. Proc Natl Acad Sci USA 96:7986–7991

Robertson KM et al (2001) Characterization of the fertility of male aromatase knockout mice. J Androl 22:825–830

Rochira V et al (2002) Congenital estrogen deficiency in men: a new syndrome with different phenotypes: clinical and therapeutic implications in men. Mol Cell Endocrinol 193:19–28

Santen RJ et al (2001) Use of the aromatase inhibitor anastrozole in the treatment of patients with advanced prostate carcinoma. Cancer 92:2095–2101

Shozu M, Simpson ER (1998) Aromatase expression of human osteoblast-like cells. Mol Cell Endocrinol 139:117–129

Simpson ER (2004) Models of aromatase insufficiency. Remin Reprod Med 22:25–30

Simpson ER, Rubin G, Clyne C et al (2000) The role of local estrogen biosynthesis in males and females. Trends Endocrinol 5:184–188

Simpson ER et al (2002) Aromatase – a brief overview. Annu Rev Physiol 64:93–127

Smith EP, Boyd J, Frank GR et al (1994) Estrogen resistance caused by a mutation in the estrogen receptor gene in a man. N Engl J Med 331:1056–1061

Smith MR et al (2003) Selective aromatase inhibition for patients with androgen-independent prostate carcinoma. Cancer 95:1864–1868

Takeda K et al (2003) Progressive development of insulin resistance phenotype in male mice with complete aromatase (CYP19) deficiency. J Endocrinol 176:237–246

Toda K, Terashima M, Kamamoto T et al (1990) Structural and functional characterization of human aromatase. Eur J Biochem 193:559–565

Toda K, Okada T, Takeda K et al (2001a) Oestrogen at the neonatal stage is critical for the reproductive ability of male mice as revealed by supplementation with 17beta-oestradiol to aromatase gene (Cyp19) knockout mice. J Endocrinol 168:455–463

Toda K, Takeda K, Akira S et al (2001b) Alternations in hepatic expression of fatty-acid metabolizing enzymes in ArKO mice and their reversal by the treatment with 17beta-estradiol or a peroxisome proliferator. J Steriod Biochem Mol Biol 79:11–17

Wagner CK, Morrell JI (1997) Neuroanatomical distribution of aromatase MRNA in the rat brain: indications of regional regulation. J Steroid Biochem Mol Biol 61:307–314

Wickman S et al (2003) Effects of suppression of estrogen action by the P450 aromatase inhibitor letrozole on bone mineral density and bone turnover in pubertal boys. J Clin Endocrinol Metab 88:3785–3793

Zabolotny BP et al (2005) Successful use of letrozole in male breast cancer: a case report and review of hormonal therapy for male breast cancer. J Surg Oncol 90:26–30

Zhou P et al (2005) Letrozole significantly improves growth potential in a pubertal boy with growth hormone deficiency. Pediatrics 115:e245–e248

Estradiol Action in Atherosclerosis and Reendothelialization

J.-F. Arnal(✉), H. Laurell, F. Lenfant, V. Douin-Echinard,
L. Brouchet, P. Gourdy

INSERM U589, INSTITUT L. BUGNARD, 1 avenue Jean Poulhès, 31403 Toulouse Cedex, France
email: *arnal@toulouse.inserm.fr*

1	The Atheromatous Process:	
	Numerous Cellular Players for Several Scenarios	71
2	E2 Prevents Fatty Streak in Models of Atheroma	73
3	Involvement of the Inflammatory–Immune System	74
4	Does E2 Have a Pro- or Anti-inflammatory Effect?	74
5	Estrogen Receptor α	
	Mediates Most of the Vascular Effects of E2	75
6	Effect on Artery Healing:	
	Reendothelialization Mechanisms	77
7	Pathophysiological and Therapeutical Implications	79
References		80

Abstract. Whereas hormonal replacement/menopause therapy (HRT) in postmenopausal women increases coronary artery disease risk, epidemiological studies (protection in premenopaused women) suggest and experimental studies (prevention of the development of fatty streaks in animals) demonstrate a major atheroprotective action of estradiol (E2). The understanding of the deleterious and beneficial effects of estrogens is thus required. The atheroprotective effect of E2 is absent in mice deficient in mature T and B lymphocytes, demonstrating the crucial role of the endothelium/immune system pair. The immunoinflammatory system appears to play a key role in the development of fatty streak deposit as well as in the rupture of the atherosclerotic plaque. Whereas E2 favors an anti-inflammatory effect in vitro (cultured cells), it elicits in vivo a proinflammation at the level of several subpopulations of the immunoinflammatory system, which could contribute to plaque destabilization. Endothelium appears to be an important target for E2, since it potentiates endothelial NO and prostacyclin production, thus promoting beneficial effects such as vasorelaxation and inhibition of platelet aggregation. Prostacyclin, but not NO, appear to be involved in the atheroprotective effect of E2, which also accelerates endothelial regrowth, thus favoring vascular healing. Finally, most of these E2 effects are mediated by estrogen receptor α and are independent of estrogen receptor β. In summary, a better understanding of the mechanisms of estrogens on the normal and atheromatous arteries is required and should help to optimize the prevention of cardiovascular disease after menopause. These mouse models should help to screen existing and future selective estrogen receptor modulators (SERMs).

Estrogens play a pivotal role in sexual development and reproduction and are also implicated in a number of physiological processes in various tissues, including the cardiovascular system. Numerous epidemiological studies suggest that estrogens protect women against cardiovascular diseases before the age of menopause. After menopause, women's cardiovascular risk becomes progressively closer to that of men, reinforcing the hypothesis of an atheroprotective effect of estrogens. However, the two controlled prospective and randomized studies published so far did not demonstrate a beneficial effect of hormone replacement therapy (HRT), neither in secondary prevention (Heart and Estrogen/Progestin Replacement Study, HERS) (Hulley et al. 1998) nor in primary prevention (Women's Health Initiative study, WHI) (Rossouw et al. 2002). This is in contrast to the large amount of data from experimental models of atherosclerosis, where estradiol (E2) treatment prevents the development of fatty streaks in comparison with castrated animals given a placebo (Hodgin and Maeda 2002).

Waters et al. (2004) recently defined five main priorities in the area of menopause treatment and cardiovascular (CV) risk including: 1. The determination of the mechanisms of the CV events during the first year of the HRT, 2. The understanding of the beneficial effects of endogenous estrogens.

1 The Atheromatous Process: Numerous Cellular Players for Several Scenarios

The first step of the atheromatous process is the penetration of atherogenic lipoproteins, in particular low-density lipoproteins (LDL) through the endothelial monolayer (Steinberg 2002). LDL oxidation occurs in the subendothelial space and probably represents a necessary modification to the subsequent steps, because oxidized LDL induces in turn an activation of the endothelium, consisting in particular of the increased expression of adhesion molecules, such as intercellular adhesion molecule (ICAM-1) and vascular cell adhesion molecule (VCAM-1). These molecules are required to slow down circulating monocytes, stop them, and allow their subsequent migration into the intima. In the subendothelial space, the activation of the monocytes induces their differentiation into macrophages, and this probably contributes to increasing the level of LDL oxidation. These modified LDLs can be recognized by scavenger receptors expressed by macrophages. Thus, macrophages attempt to clean the intima, thereby preventing the accumulation of oxidized LDL. As oxidized LDL accumulates intracellularly, macrophages progressively turn into foam cells, which make up the major component of fatty streaks.

Blood flow shear stress represents a crucial protective factor, where abnormal shear stress promotes endothelial activation and dysfunction, to date the best-recognized explanation for a focal location of atheroma functions (Gimbrone 1999; Shyy and Chien 2002; Tedgui and Mallat 2001). Classic risk factors (high blood pressure, hypercholesterolemia, smoking, diabetes) appear to favor and/or aggravate endothelial dysfunction. They can also favor the production of chemokines and cytokines by the different cellular actors (endothelium, monocytes-macrophages, smooth muscle cells, but also lymphocytes). Protective factors are less recognized, although high-density lipoproteins (HDL) appear to be of major importance.

Expansion of the fatty streak tends to be limited and circumvented by a scarring reaction of the smooth muscle cells migrating to the intima and secreting collagen. The balance between the inflammation level and the strength of fibromuscular cap determines the stability of the atheromatous plaque (Ross 1999). Plaque rupture exposes thrombogenic materials, leading to the formation of a thrombus, which threatens the viability of the tissue downstream from the occluded artery. Unfortunately, plaque rupture is not satisfactorily modeled in mouse models, and this is probably the greatest limitation of the current experimental approach.

Many groups have been working to describe the cellular and molecular mechanisms leading to the aggravation of or protection from atheroma (Ross 1999; Tedgui and Mallat 2001; Hansson et al. 2002; Libby 2002), leading to the generation of the two major models of hypercholesterolemic mice: mice deficient in apolipoprotein E (apoE-KO) and mice deficient in LDL-receptor. ApoE-KO mice are hypercholesterolemic (3–4 g cholesterol/l) under a chow diet and have very low levels of HDL cholesterol (Zhang et al. 1992). Accordingly, they spontaneously and rapidly (within a few weeks) develop fatty streak at the root of the aorta. Mice deficient in LDL-receptor must be given a Western diet (fat plus cholesterol) to develop fatty streaks, because their lipoprotein profile under chow diet is less severe than the apoE-KO mice.

So far, the cellular and molecular dissection of the pathophysiology of atheroma was explored by breeding hypercholesterolemic mice and mice deficient in another specific gene. For instance, hypercholesterolemic mice also deficient in either monocyte-macrophage (through a deficit of macrophage colony stimulating factor) (Smith et al. 1995) or mature B and T lymphocytes (*RAG-2* gene deficient) (Dansky et al. 1997) respectively develop one-tenth and one-half the amount of fatty streaks than control hypercholesterolemic mice. Mice deficient in various cytokines in general demonstrated the aggravating role of pro-inflammatory cytokine (such as interferon γ (IFNγ), interleukins (IL)-1α and β, IL-12, IL-18) and the protective role of anti-inflammatory cytokines (mainly IL-10) in the development of the atherosclerotic process (see Libby 2002 and Tedgui and Mallat 2001 and references therein).

Platelets were recently found to participate to the constitution of fatty streak lesions at a very early stage, in particular at the level of the carotid bifurcation, a lesion-prone site, by interacting with the activated endothelium before any macrophage infiltration (Massberg et al. 2002). This process involves the platelet GPIbα (glycoprotein 1b, alpha polypeptide), and the adhesive proteins P-selectin and/or von Willebrand factor, which mediate the attachment of platelets to activated endothelial cells. Blocking these interactions completely (–100%) prevented fatty streak formation at the level of the carotid bifurcation in apoE-KO mice, but only partially (−30%) at the level of the aortic sinus (Massberg et al. 2002). While this points out that platelets can be a target for antiatherosclerotic therapies, this also suggests that modulation of the adhesive properties of the endothelium may be of pathophysiological relevance, especially at the level of the carotid bifurcation.

2 E2 Prevents Fatty Streak in Models of Atheroma

Studies in primates, mainly conducted by Clarkson et al., have provided convincing evidence for the primary prevention of coronary artery atherosclerosis when estrogens are administered soon after the development of estrogen deficiency (Clarkson and Appt 2005). Equally convincing are the data from monkey studies indicating the total loss of the beneficial effects of estrogens if the treatment is delayed for a period equal to 6 postmenopausal years for women. Moreover, in the monkey model, an attempt has been made to identify the most effective hormone treatment regimen in preventing the progression of coronary artery atherosclerosis. By far the most successful approach uses estrogen containing oral contraceptive during the perimenopausal transition, followed directly by hormone replacement therapy postmenopausally. However, the monkey model does not provide an understanding of the cellular or molecular mechanisms of E2 action.

Several groups, including ours, have been working to describe the vascular effects of E2 and elucidate the cellular and molecular mechanisms. Castration of ApoE-KO or LDL receptor (LDL-R) KO mice is followed by an increase in the fatty streak lesion area and, in castrated mice, E2 prevents this fatty streak deposit. However, serum E2 concentrations on the order of those encountered during gestation are necessary for maximal protection (Bourassa et al. 1996; Elhage et al. 1997a). Although E2 treatment induces a decrease in serum cholesterol concentrations, the decrease involving both LDL and HDL fractions is too minor to explain the hormone's atheroprotective effect (Elhage et al. 1997a; Hodgin and Maeda 2002), which seems to result mainly from a direct effect of E2 on the cells of the arterial wall (Arnal et al. 2004). A similar conclusion was previously obtained by other groups (Haarbo et al. 1991; Holm et al. 1999) using hypercholesterolemic rabbits. In addition, they elegantly showed the crucial role of intact endothelium, as the antiatherogenic effect of E2 was abolished, or even reversed, after balloon catheter injury (Holm et al. 1999).

3 Involvement of the Inflammatory–Immune System

A mentioned above, cell populations of the inflammatory–immune system (monocytes-macrophages, lymphocytes, etc.) play crucial roles in the pathophysiology of atherosclerosis (Binder et al. 2002; Hansson et al. 2002; Libby 2002; Tedgui and Mallat 2001). Indeed, we demonstrated that E2 prevents the deposit of fatty streaks in immunocompetent ApoE-KO mice, whereas it has no effect in mice deficient in both apoE and *RAG-2* gene expression, since mature B and T lymphocytes are lacking (Elhage et al. 2000). One hypothesis resulting from these observations was that lymphocytes, or at least a subpopulation of them, were the mediators of the atheroprotective effect. After crossing ApoE-KO mice with mice deficient in either TCR$\alpha\beta$, CD4, CD8, or TCR$\delta\gamma$T lymphocytes, we reported that TCR$\alpha\beta$ T lymphocytes play a major role in fatty streak development (Elhage et al. 2004). However, the protective effect of E2 persisted in all these strains, showing that none of these lymphocyte subpopulations specifically mediated the atheroprotective effect of E2 (Elhage et al. 2005).

4 Does E2 Have a Pro- or Anti-inflammatory Effect?

At variance with macrophages in fatty streaks, peritoneal macrophages are a cell population that can be obtained in considerable amounts, and thus the chronic effect of E2 on the inflammatory–immune system can easily be studied in these cells. We first must mention again that a chronic in vivo treatment of mice by E2 led to a pro-inflammatory response in peritoneal macrophages, whereas an acute (only a few hours of E2) in vitro treatment of the macrophage cell line RAW 264.7 by E2 led to an anti-inflammatory effect (Ghisletti et al. 2005). This striking discrepancy demonstrates the importance of the in vivo approach to understand the pathophysiological effects of E2.

We also evaluated the possibility that E2 promotes the production of anti-inflammatory cytokines. In collaboration with two groups of immunologists, we demonstrated that the profile of cytokine secretion in CD4$^+$ (Maret et al. 2003) as well as in NK T lymphocytes (Gourdy et al. 2005) is altered by E2. In these studies, an increase in IFNγ production and a decrease in anti-inflammatory cytokine production were observed, i.e., toward a Th1 profile. Similarly, we observed an increased production of IL-1 (α and β), IL-12, and IL-18 by macrophages obtained from E2-treated mice compared with those from castrated mice (B. Calippe et al., unpublished data).

According to current knowledge, the pro-inflammatory effect of E2 cannot account for its preventive effect of fatty streak accumulation (Binder et al. 2002; Hansson et al. 2002; Libby 2002; Tedgui and Mallat 2001). In contrast, it could contribute to the destabilization of atheromatous plaques, and thus represent a good candidate to explain the increase in cardiovascular events during the year following the onset of HRT (Hulley et al. 1998; Rossouw et al. 2002). Nonetheless, the non-natural progestin (medroxyprogesterone acetate) used in these clinical trials could indeed have undesirable, deleterious effects (Anderson et al. 2004).

5 Estrogen Receptor α Mediates Most of the Vascular Effects of E2

Endothelium is involved in the regulation of coagulation, leukocyte adhesion in inflammation, transvascular flux of cells, vascular smooth muscle growth, etc. and also is a major target for E2. Endothelial messengers, such as nitric oxide (NO) (Mendelsohn 2000 and references therein) and prostacyclin are increased by E2. Indeed, E2 can increase NO bioactivity acutely by stimulating endothelial NO synthase activity (Mendelsohn 2000) and chronically by decreasing the breakdown of NO, as a consequence of a decreased production of reactive oxygen species (Arnal et al. 1996; Wagner et al. 2001). Although the atheroprotective effect of E2 appears independent of NO production (Elhage et al. 1997b), induction of COX-2 and prostacyclin production was recently proposed to play an important role in the prevention of fatty streak at the level of the thoracoabdominal aorta (Egan et al. 2004).

E2 effects can be mediated by estrogen receptor alpha (ERα) and beta (ERβ), two members of the nuclear receptor superfamily that are encoded by two distinct genes (Couse and Korach 1999). A collaborative effort with the Krust and Chambon group led to the clearcut demonstration of a prominent role of ERα in vascular physiology in vivo. Full length ERα (66 kDa) is composed of six domains (named from A to F) and two independent transactivation functions AF1 and AF2 (Tora et al. 1989). The ERα and ERβ genes were previously disrupted by targeted mutagenesis (Couse and Korach 1999). Mice targeted for ERα through the insertion of the *Neo* gene in exon 1 (therefore named ER-αNeoKO) was first generated by Korach's group in 1993 (Lubahn et al. 1993). These mice were subsequently shown to present a leakage due to a non-natural alternative splicing of the ERα mRNA, resulting in the expression of a truncated 55-kD isoform (Couse et al. 1995; Kos et al. 2002; Pendaries et al. 2002). Such an ERα isoform, lacking most of the A/B domain and therefore the AF-1 transactivating function, was sufficient to mediate the E2 effect on the endothelial NO production (Pendaries et al. 2002) and postinjury medial hyperplasia (Iafrati et al. 1997; Pare et al. 2002).

In contrast, the generation and studies of mice that fully and unambiguously lack ERα (Dupont et al. 2000) showed that ERα is necessary in the response of E2 on NO production (Pendaries et al. 2002). Erβ-deficient mice had, however, a normal NO production (Darblade et al. 2002). Altogether, these data allow us to conclude that an ERα lacking the A/B domain (and therefore AF-1) is sufficient to mediate some of the vascular effects of estrogen. Interestingly, an ERα 46-kD isoform, lacking the N-terminal portion (domains A/B), can be physiologically expressed through an alternative splicing (Flouriot et al. 2000) in the uterus (Faye et al. 1986; Pendaries et al. 2002), in cultured endothelial cells (Li et al. 2003) and in macrophages (H. Laurell, unpublished data).

However, the prevention of fatty streaks appears to require the full length ERα (66 kD) (Hodgin et al. 2001). Indeed, in contrast to the E2 protection elicited in apoE-KO mice, E2 treatment of ovariectomized ERα-Neo/apoE double KO female mice caused a nonsignificant ($p = 0.12$) reduction in lesion size and no reduction in total plasma cholesterol (Hodgin et al. 2001). However, it should be mentioned that E2 treatment significantly reduced the complexity of plaques in ERα-Neo/apoE double KO female mice, although not to the same degree as in apoE-KO female mice. Although this could have been due to the existence of ERα-dependent atheroprotective effects of E2, the expression of the truncated 55kD ER-α isoform could also have been responsible for this E2 effect.

6 Effect on Artery Healing: Reendothelialization Mechanisms

As mentioned in the introduction, the loss of the integrity of the endothelial monolayer is another important aspect at early steps of atherosclerosis, but also after the destruction provoked by endoluminal angioplasty (often followed by stent implantation) (Bennett and O'Sullivan 2001). In this context, the acceleration of vascular healing, where reendothelialization plays a key role, is considered a major protective event against short- as well as long-term complications of endovascular therapy.

Endovascular de-endothelialization in mice is a complex and delicate manipulation as a consequence of the very small size of the carotid artery. Carmeliet et al. (1997) proposed destroying the endothelium using a perivascular electric injury approach, and an adaptation of this model was proposed by our group (Brouchet et al. 2001). Endovascular and electric perivascular injury are identical in their efficiency in destroying the endothelium, which will be temporarily replaced by a monolayer of platelets, but they differ in at least two major points. Firstly, whereas endovascular injury preserves most of the medial smooth muscle cells as well as adventitia cells (Lindner et al. 1993), electric injury destroys the cells of the three layers of the injured area, and in particular the smooth muscle cells, which do not recolonize the media even after several weeks. Secondly, immunoinflammatory cells (mainly macrophages) make up the main cell population in the perivascular electric injury model that may interact with regenerating endothelium. The accelerative effect of E2 on reendothelialization is mediated by ERα (Brouchet et al. 2001) and endothelial NO synthase appears absolutely required for this effect (Iwakura et al. 2003). We have recently explored the role of two molecules that also appear to be clearly involved in this process: fibroblast growth factor 2 (FGF2) and osteopontin (OPN).

FGF2 is one of the first growth factors to be characterized and remains one of the most potent. The expression of the five FGF2 isoforms of 18, 22, 22.5, 24, and 34 kDa in humans and the three isoforms of 18, 21, and 22 kDa in mice is particularly original, since they are synthesized not through alternative splicing of mRNA, but through an alternative use of translation initiation codons from a single mRNA (Prats et al. 1989). These isoforms differ only in their NH2 extremities, which confer a nuclear localization to the high molecular weight (HMW) CUG-initiated forms, whose function is for the most part unknown. In contrast, the low-molecular-weight (LMW) AUG-initiated FGF2 (18 kDa) is predominantly cytoplasmic and excreted, and known to activate the FGF receptors (FGFR), leading to stimulation of proliferation and migration. We recently reported that E2 stimulated migration in endothelial cells from $Fgf2^{+/+}$, but not from $Fgf2^{-/-}$ mice (deficient both in FGF2^{lmw} and in FGF2^{hmw}) (Garmy-Susini et al. 2004). More recently, we confirmed that E2 increased both the velocity of reendothelialization and the number of circulating EPCs (as previously described in Strehlow et al. 2003) in ovariectomized $Fgf2^{+/+}$ mice. However, both these effects of E2 were abolished in $Fgf2^{-/-}$ mice. We then investigated the role of medullary FGF2 in these processes. In chimeric (i.e., bone-marrow [BM] transplanted) ($Fgf2^{-/-}$ BM $\Rightarrow Fgf2^{+/+}$) mice, both effects of E2 on reendothelialization and on circulating EPC levels were abolished, whereas both were preserved in chimeric ($Fgf2^{+/+}$ BM $\Rightarrow Fgf2^{-/-}$) mice, demonstrating that medullary, and not extramedullary, FGF2 is required for both effects of E2 (Fontaine et al. 2006). Similarly, ERα$^{+/+}$ or ERα$^{-/-}$ BM graft experiments revealed that the effect of E2 on reendothelialization relies on medullary, but not on extramedullary ERα, emphasizing for the first time, to the best of our knowledge, a prominent role played by bone marrow in the E2 effect on reendothelialization.

OPN is an RGD-containing extracellular matrix phosphoprotein involved in cell adhesion and migration via a number of receptors, including several integrins and CD44 (Chaulet et al. 2001). $OPN^{-/-}$ mice are resistant to ovariectomy-induced bone resorption (Yoshitake et al.

1999). OPN is detected in vascular SMCs and macrophages of atherosclerotic plaques (Giachelli et al. 1993) and its expression is upregulated in neointimal hyperplasia and in regenerating endothelium (Liaw et al. 1995). However, the cellular source and target of OPN in accelerating endothelial regeneration remains to be demonstrated. In collaboration with Gadeau et al., we recently found that the effect de E2 on reendothelialization was abolished in $OPN^{-/-}$ mice in the two models of carotid injury (unpublished data).

7 Pathophysiological and Therapeutical Implications

In conclusion, E2 exerts an atheroprotective effect in all experimental models and most likely in women before menopause. Although serum cholesterol decreases, this influence on lipid metabolism is negligible. Similarly, although E2 induces an increase in endothelial NO production and/or bioavailability, this effect does not account for its protection of the constitution of fatty streak. The precise mechanisms of the atheroprotective effect of E2 at the level of the endothelium remains to be characterized. At the same time, E2 also induces an immunoinflammatory response toward a Th1 profile with increasing interferon γ production. This proinflammatory effect could have been prominent in advanced atheromatous plaques in postmenopausal women, favoring destabilization of the most unstable plaques. At present, this is the most likely explanation accounting for the increase in the frequency of cardiovascular events in postmenopausal women during the 1st year of HRT, as observed in the HERS and WHI studies. It should be noted that the women enrolled in these studies had been postmenopausal for several years (on average more than 10 years after the onset of menopause).

Although ERs are classically defined as ligand-activated transcription factors, it has become clear that extragenomic membrane short-term responses play an important role in cultured endothelial cells (Mendelsohn 2000), but also in osteoblasts (such as the activation of PI3kinases-AKT pathways as well as MAP kinase pathways) (Kousteni et al. 2001). An important challenge for the next years will be to describe the respective roles of these membrane effects and the classic effects.

These new acquisitions are a basis for new pharmacological developments that can prevent harmful effects and preserve the beneficial effects. The effects of selective estrogen receptor modulators (SERMs) on the different constituents in the atheroma plaque formation must now be analyzed on the basis of their specific regulation of the ERα but also of the ERβ. Various classes of estrogens and SERMs have been described according their molecular actions through ERα (Jordan 2001a,b; Katzenellenbogen and Katzenellenbogen 2002).

Due to the complexity of the mechanisms of action of estrogens and SERMs, their effect on each type of cell and tissue cannot be predicted from their structure. Indeed, only integrated models can screen the present and future SERMs in terms of the beneficial and deleterious effects. Theoretically, it is conceivable to design a SERM (or a combination of molecules) that retains most (if not all) of the desired effects of E2 (on the central nervous system to prevent hot flushes, on bone, endothelium, etc.), but which is devoid of the undesirable effects of E2 (mainly breast cancer and thromboembolism). Finally, though not within the scope of this review, it should be mentioned that the interaction with progestins is of major importance and should also be studied in parallel investigations.

Acknowledgements. We are grateful to Prof. F. Bayard for his input to our team's work over many years. We are grateful to P. Chambon, A. Krust, K. Korach, J.C. Guery, A.P. Gadeau, C. Filipe, A. Billon, and B. Calippe for helpful discussions over many years. We thank M.J. Fouque and A. Schambourg for their skillful technical assistance. The work described herein was supported in part by Université Paul Sabatier Toulouse III, INSERM, the European Vascular Genomics Network No. 503254, the Fondation de France, the Fondation de l'Avenir, and the Conseil Régional Midi-Pyrénées in France.

References

Anderson GL, Limacher M, Assaf AR, Bassford T, Beresford SA, Black H, Bonds D, Brunner R, Brzyski R, Caan B et al (2004) Effects of conjugated equine estrogen in postmenopausal women with hysterectomy: the Women's Health Initiative randomized controlled trial. JAMA 291:1701–1712

Arnal JF, Clamens S, Pechet C, Nègre-Salvayre A, Allera C, Girolami JP, Salvayre R, Bayard F (1996) Ethinylestradiol does not enhance the expression of nitric oxide synthase in bovine aortic endothelial cells but increases the release of bioactive nitric oxide by inhibiting superoxide anion production. Proc Natl Acad Sci USA 93:4108–4113

Arnal JF, Gourdy P, Elhage R, Garmy-Susini B, Delmas E, Brouchet L, Castano C, Barreira Y, Couloumiers JC, Prats H et al (2004) Estrogens and atherosclerosis. Eur J Endocrinol 150:113–117

Bennett MR, O'Sullivan M (2001) Mechanisms of angioplasty and stent restenosis: implications for design of rational therapy. Pharmacol Ther 91:149–166

Binder CJ, Chang MK, Shaw PX, Miller YI, Hartvigsen K, Dewan A, Witztum JL (2002) Innate and acquired immunity in atherogenesis. Nat Med 8:1218–1226

Bourassa PA, Milos PM, Gaynor BJ, Breslow JL, Aiello RJ (1996) Estrogen reduces atherosclerotic lesion development in apolipoprotein E-deficient mice. Proc Natl Acad Sci USA 93:10022–10027

Brouchet L, Krust A, Dupont S, Chambon P, Bayard F, Arnal JF (2001) Estradiol accelerates reendothelialization in mouse carotid artery through estrogen receptor-alpha but not estrogen receptor-beta. Circulation 103:423–428

Carmeliet P, Moos L, Stassen J, De Mol M, Bouché A, van den Oord J, Kocks M, Collen D (1997) Vascular wound healing and neointima formation induced by perivascular electric injury in mice. Am J Pathol 150:761–776

Chaulet H, Desgranges C, Renault MA, Dupuch F, Ezan G, Peiretti F, Loirand G, Pacaud P, Gadeau AP (2001) Extracellular nucleotides induce arterial smooth muscle cell migration via osteopontin. Circ Res 89:772–778

Clarkson TB, Appt SE (2005) Controversies about HRT: lessons from monkey models. Maturitas 51:64–74

Couse JF, Curtis SW, Washburn TF, Lindzey J, Golding TS, Lubahn DB, Smithies O, Korach KS (1995) Analysis of transcription and estrogen insensitivity in the female mouse after targeted disruption of the estrogen receptor gene. Mol Endocrinol 9:1441–1454

Couse JF, Korach KS (1999) Estrogen receptor null mice: what have we learned and where will they lead us? Endocr Rev 20:358–417

Dansky H, Charlton S, McGee Harper M, Smith J (1997) T and B lymphocytes play a minor role in atherosclerotic plaque formation in the apolipoprotein E-deficient mouse. Proc Natl Acad Sci USA 94:4642–4646

Darblade B, Pendaries C, Krust A, Dupont S, Fouque MJ, Rami J, Chambon P, Bayard F, Arnal JF (2002) Estradiol alters nitric oxide production in the mouse aorta through the alpha-, but not beta-, estrogen receptor. Circ Res 90:413–419

Dupont S, Krust A, Gansmuller A, Dierich A, Chambon P, Mark M (2000) Effect of single and compound knockouts of estrogen receptors alpha (ERalpha) and beta (ERbeta) on mouse reproductive phenotypes. Development 127:4277–4291

Egan KM, Lawson JA, Fries S, Koller B, Rader DJ, Smyth EM, Fitzgerald GA (2004) COX-2-derived prostacyclin confers atheroprotection on female mice. Science 306:1954–1957

Elhage R, Arnal JF, Pierragi M-T, Duverger N, Fiévet C, Faye JC, Bayard F (1997a) Estradiol-17b prevents fatty streak formation in apolipoprotein E-deficient mice. Arterioscl Thromb Vasc Biol 17:2679–2684

Elhage R, Bayard F, Richard V, Holvoet P, Duverger N, Fievet C, Arnal JF (1997b) Prevention of fatty streak formation of 17beta-estradiol is not mediated by the production of nitric oxide in apolipoprotein E-deficient mice. Circulation 96:3048–3052

Elhage R, Clamens S, Reardon-Alulis C, Getz GS, Fievet C, Maret A, Arnal JF, Bayard F (2000) Loss of atheroprotective effect of estradiol in immunodeficient mice. Endocrinology 141:462–465

Elhage R, Gourdy P, Brouchet L, Jawien J, Fouque MJ, Fievet C, Huc X, Barreira Y, Couloumiers JC, Arnal JF, Bayard F (2004) Deleting TCR alpha beta+ or CD4+ T lymphocytes leads to opposite effects on site-specific atherosclerosis in female apolipoprotein E-deficient mice. Am J Pathol 165:2013–2018

Elhage R, Gourdy P, Huc X, Brouchet L, Castano C, Barreira Y, Couloumiers JC, Fievet C, Hansson G, Arnal J-F et al (2005) The atheroprotective effect of 17-estradiol depends on complex interactions in adaptive immunity. Am J Pathol 167:267–274

Faye JC, Fargin A, Bayard F (1986) Dissimilarities between the uterine estrogen receptor in cytosol of castrated and estradiol-treated rats. Endocrinology 118:2276–2283

Flouriot G, Brand H, Denger S, Metivier R, Kos M, Reid G, Sonntag-Buck V, Gannon F (2000) Identification of a new isoform of the human estrogen receptor-alpha (hER-alpha) that is encoded by distinct transcripts and that is able to repress hER-alpha activation function 1. EMBO J 19:4688–4700

Fontaine V, Filipe C, Werner N, Gourdy P, Billon A, Garmy-Susini B, Bouchet L, Bayard F, Prats H, Doetschman T, Nickenig G, Arnal JF (2006) Essential role of bone marrow fibroblast growth factor-2 in the effect of estradiol on reendothelialization and endothelial progenitor cell mobilization. Am J Pathol 169:1855–1862

Garmy-Susini B, Delmas E, Gourdy P, Zhou M, Bossard C, Bugler B, Bayard F, Krust A, Prats AC, Doetschman T et al (2004) Role of fibroblast growth factor-2 isoforms in the effect of estradiol on endothelial cell migration and proliferation. Circ Res 94:1301–1309

Ghisletti S, Meda C, Maggi A, Vegeto E (2005) 17{beta}-Estradiol inhibits inflammatory gene expression by controlling NF-{kappa}B intracellular localization. Mol Cell Biol 25:2957–2968

Giachelli CM, Bae N, Almeida M, Denhardt DT, Alpers CE, Schwartz SM (1993) Osteopontin is elevated during neointima formation in rat arteries and is a novel component of human atherosclerotic plaques. J Clin Invest 92:1686–1696

Gimbrone MA Jr (1999) Vascular endothelium, hemodynamic forces, and atherogenesis. Am J Pathol 155:1–5

Gourdy P, Araujo LM, Zhu R, Garmy-Susini B, Diem S, Laurell H, Leite-de-Moraes M, Dy M, Arnal JF, Bayard F, Herbelin A (2005) Relevance of sexual dimorphism to regulatory T cells: estradiol promotes IFN-{gamma} production by invariant natural killer T cells. Blood 105:2415–2420

Haarbo J, Leth-Espensen P, Stender S, Christiansen C (1991) Estrogen monotherapy and combined estrogen-progestogen replacement therapy attenuate aortic accumulation of cholesterol in ovariectomized cholesterol-fed rabbits. J Clin Invest 87:1274–1279

Hansson GK, Libby P, Schonbeck U, Yan ZQ (2002) Innate and adaptive immunity in the pathogenesis of atherosclerosis. Circ Res 91:281–291

Hodgin J, Maeda N (2002) Estrogen and mouse models of atherosclerosis. Endocrinology 143:4495–4501

Hodgin JB, Krege JH, Reddick RL, Korach KS, Smithies O, Maeda N (2001) Estrogen receptor alpha is a major mediator of 17beta-estradiol's atheroprotective effects on lesion size in Apoe$^{-/-}$ mice. J Clin Invest 107:333–340

Holm P, Andersen HL, Andersen MR, Erhardtsen E, Stender S (1999) The direct antiatherogenic effect of estrogen is present, absent, or reversed, depending on the state of the arterial endothelium. A time course study in cholesterol-clamped rabbits. Circulation 100:1727–1733

Hulley S, Grady D, Bush T, Furberg C, Herrington D, Riggs B, Vittinghoff E (1998) Randomized trial of estrogen plus progestin for secondary prevention of coronary heart disease in postmenopausal women. Heart and Estrogen/progestin Replacement Study (HERS) Research Group. JAMA 280:605–613

Iafrati MD, Karas RH, Aronovitz M, Kim S, Sullivan TR Jr, Lubahn DB, O'Donnell TF Jr, Korach KS, Mendelsohn ME (1997) Estrogen inhibits the vascular injury response in estrogen receptor alpha-deficient mice. Nat Med 3:545–548

Iwakura A, Luedemann C, Shastry S, Hanley A, Kearney M, Aikawa R, Isner JM, Asahara T, Losordo DW (2003) Estrogen-mediated, endothelial nitric oxide synthase-dependent mobilization of bone marrow-derived endothelial progenitor cells contributes to reendothelialization after arterial injury. Circulation 108:3115–3121

Jordan VC (2001a) The past, present, and future of selective estrogen receptor modulation. Ann NY Acad Sci 949:72–79

Jordan VC (2001b) Selective estrogen receptor modulation: a personal perspective. Cancer Res 61:5683–5687

Katzenellenbogen BS, Katzenellenbogen JA (2002) Biomedicine. Defining the "S" in SERMs. Science 295:2380–2381

Kos M, Denger S, Reid G, Korach KS, Gannon F (2002) Down but not out? A novel protein isoform of the estrogen receptor alpha is expressed in the estrogen receptor alpha knockout mouse. J Mol Endocrinol 29:281–286

Kousteni S, Bellido T, Plotkin LI, O'Brien CA, Bodenner DL, Han L, Han K, DiGregorio GB, Katzenellenbogen JA, Katzenellenbogen BS et al (2001) Nongenotropic, sex-nonspecific signaling through the estrogen or androgen receptors: dissociation from transcriptional activity. Cell 104:719–730

Li L, Haynes MP, Bender JR (2003) Plasma membrane localization and function of the estrogen receptor alpha variant (ER46) in human endothelial cells. Proc Natl Acad Sci USA 100:4807–4812

Liaw L, Lindner V, Schwartz SM, Chambers AF, Giachelli CM (1995) Osteopontin and beta 3 integrin are coordinately expressed in regenerating endothelium in vivo and stimulate Arg-Gly-Asp-dependent endothelial migration in vitro. Circ Res 77:665–672

Libby P (2002) Inflammation in atherosclerosis. Nature 420:868–874

Lindner V, Fingerle J, Reidy MA (1993) Mouse model of arterial injury. Circ Res 73:792–796

Lubahn DB, Moyer JS, Golding TS, Couse JF, Korach KS, Smithies O (1993) Alteration of reproductive function but not prenatal sexual development after insertional disruption of the mouse estrogen receptor gene. Proc Natl Acad Sci USA 90:11162–11166

Maret A, Coudert JD, Garidou L, Foucras G, Gourdy P, Krust A, Dupont S, Chambon P, Druet P, Bayard F, Guery JC (2003) Estradiol enhances primary antigen-specific CD4 T cell responses and Th1 development in vivo. Essential role of estrogen receptor alpha expression in hematopoietic cells. Eur J Immunol 33:512–521

Massberg S, Brand K, Gruner S, Page S, Muller E, Muller I, Bergmeier W, Richter T, Lorenz M, Konrad I et al (2002) A critical role of platelet adhesion in the initiation of atherosclerotic lesion formation. J Exp Med 196:887–896

Mendelsohn ME (2000) Nongenomic ER-mediated activation of endothelial nitric oxide synthase: how does it work? What does it mean? Circ Res 87:956–960

Pare G, Krust A, Karas RH, Dupont S, Aronovitz M, Chambon P, Mendelsohn ME (2002) Estrogen receptor-alpha mediates the protective effects of estrogen against vascular injury. Circ Res 90:1087–1092

Pendaries C, Darblade B, Rochaix P, Krust A, Chambon P, Korach KS, Bayard F, Arnal JF (2002) The AF-1 activation-function of ERalpha may be dispensable to mediate the effect of estradiol on endothelial NO production in mice. Proc Natl Acad Sci USA 99:2205–2210

Prats H, Kaghad M, Prats AC, Klagsbrun M, Lelias JM, Liauzun P, Chalon P, Tauber JP, Amalric F, Smith JA, Caput D (1989) High molecular mass forms of basic fibroblast growth factor are initiated by alternative CUG codons. Proc Natl Acad Sci USA 86:1836–1840

Ross R (1999) Atherosclerosis-an inflammatory disease. N Engl J Med 340: 115–125

Rossouw JE, Anderson GL, Prentice RL, LaCroix AZ, Kooperberg C, Stefanick ML, Jackson RD, Beresford SA, Howard BV, Johnson KC et al (2002) Risks and benefits of estrogen plus progestin in healthy postmenopausal women: principal results from the Women's Health Initiative randomized controlled trial. JAMA 288:321–333

Shyy JY, Chien S (2002) Role of integrins in endothelial mechanosensing of shear stress. Circ Res 91:769–775

Smith JD, Trogan E, Ginsberg M, Grigaux C, Tian J, Miyata M (1995) Decreased atherosclerosis in mice deficient in both macrophage colony-stimulating factor (op) and apolipoprotein E. Proc Natl Acad Sci USA 92:8264–8268

Steinberg D (2002) Atherogenesis in perspective: hypercholesterolemia and inflammation as partners in crime. Nat Med 8:1211–1217

Strehlow K, Werner N, Berweiler J, Link A, Dirnagl U, Priller J, Laufs K, Ghaeni L, Milosevic M, Bohm M, Nickenig G (2003) Estrogen increases bone marrow-derived endothelial progenitor cell production and diminishes neointima formation. Circulation 107:3059–3065

Tedgui A, Mallat Z (2001) Anti-inflammatory mechanisms in the vascular wall. Circ Res 88:877–887

Tora L, White J, Brou C, Tasset D, Webster N, Scheer E, Chambon P (1989) The human estrogen receptor has two independent nonacidic transcriptional activation functions. Cell 59:477–487

Wagner AH, Schroeter MR, Hecker M (2001) 17beta-Estradiol inhibition of NADPH oxidase expression in human endothelial cells. FASEB J 15:2121–2130

Waters DD, Gordon D, Rossouw JE et al (2004) Women's Ischemic Syndrome Evaluation: current status and future research directions: report of the National Heart, Lung and Blood Institute workshop: October 2–4, 2002: Section 4: lessons from hormone replacement trials. Circulation 17 109(6):e53–e55

Yoshitake H, Rittling SR, Denhardt DT, Noda M (1999) Osteopontin-deficient mice are resistant to ovariectomy-induced bone resorption. Proc Natl Acad Sci USA 96:8156–8160

Zhang SH, Reddick RL, Piedrahita JA, Maeda N (1992) Spontaneous hypercholesterolemia and arterial lesions in mice lacking apolipoprotein E. Science 258:468–471

… *Functional Effects and Molecular Mechanisms of Subtype-Selective ERα and ERβ Agonists in the Cardiovascular System*

P.A. Arias-Loza, V. Jazbutyte, K.-H. Fritzemeier,
C. Hegele-Hartung, L. Neyses, G. Ertl, T. Pelzer(✉)

Medizinische Klinik I, University of Würzburg, Josef-Schneider Str. 2, 97080 Würzburg, Germany
email: *pelzer_r@klinik.uni-wuerzburg.de*

1	Introduction	88
2	Studies of Isotype Selective ER Agonists in Spontaneously Hypertensive Rats	90
3	Studies of Isotype Selective ER Agonists in Animal Models of Aldosterone-Salt-Induced Cardiovascular Injury	95
4	Summary and Perspectives	102
	References	102

Abstract. Gender differences in the development of cardiovascular disease suggested for a protective function of estrogens in heart disease. The negative or neutral outcome of clinical trials on hormone replacement therapy provides clear evidence that the role of female sex hormones in the cardiovascular system is more complex than previously thought. In particular, the function of estrogens can not be understood without detailed knowledge on the specific function of both estrogen receptor subtypes in the heart and in the vasculature. In here, we review recent studies on subtype selective ERα and ERβ agonists in different animal models of hypertension, cardiac hypertrophy and vascular inflammation. The results indicate that the activation of specific ER subtypes confers specific as well as redundant protective effects in hypertensive heart disease that might ultimately translate into novel treatment options for hypertensive heart disease.

1 Introduction

Gender differences in the development of cardiovascular disease, which were first reported in population-based clinical trials, served as a starting point to explore the possibility that female sex hormones might serve as an important co-variable in cardiovascular health (Stampfer et al. 1991). From these studies, it became clear that the overall risk for heart disease is significantly lower for premenopausal women compared to age-matched men with an otherwise identical array of cardiovascular risk factors. The protective function of female gender is, however, rapidly lost following menopause and postmenopausal women face at least an identical risk for cardiovascular events as men at identical age (Barrett-Connor and Bush 1991; Grodstein and Stampfer 1995). The association of decreasing estrogen serum levels and increased cardiovascular disease suggested for a protective function of estrogens against heart disease (Stampfer et al. 1991). Accordingly, hopes have been raised that hormone replacement therapy might provide an additional route for the primary or secondary prevention of coronary artery disease in postmenopausal women.

The neutral or negative outcome of several prospective clinical trials such as HERS and the WHI studies have challenged these hypotheses, with the result that HRT should no longer be initiated for the sole purpose to treat or to prevent cardiovascular disease in postmenopausal women (Grady et al. 2002; Herrington and Howard 2003; Hulley et al. 1998; Rossouw et al. 2002). Beyond that, the HERS and WHI results also indicate that our current knowledge and understanding of sex hormone function in the heart is currently not sufficient to develop safer and more efficient treatment strategies for patients who might otherwise profit from HRT.

Several nuclear sex hormone receptors, including estrogen, androgen, and progesterone receptors, are functionally expressed in most cardiovascular cell types, including cardiac myocytes, vascular smooth muscle cells, endothelial cells, and fibroblasts (Grohe et al. 1997; Karas et al. 2001b). Initially, it was thought that the effects of estrogens are transmitted only by a single estrogen receptor that was cloned in the lab of P. Chambon (Green et al. 1986). The "classical" estrogen receptor was consequently termed ERα upon cloning of a second and distinct estrogen receptor, which was termed ERβ (Kuiper et al. 1996). Both ER subtypes function as ligand-dependent transcription factors and bind with high affinity to conserved *cis*-acting elements in the promoter region of estrogen responsive genes (Kuiper et al. 1997). Both ER subtypes share considerable domain structure homology, but the structure of the ligand-binding domain as well as the tissue-specific expression pattern of ERα and ERβ are very distinct and suggest that both receptors may have specific as well as redundant functions (Lindberg et al. 2003). This hypothesis has been confirmed in numerous studies conducted in transgenic mice that were independently generated by several groups and which lack either ERα, ERβ, or ERα and ERβ (Lubahn et al. 1993; Krege et al. 1998). The cardiovascular phenotype of ERKO and BERKO mice indicates that ERβ plays a relevant role in blood pressure regulation in mice because BERKO mice are hypertensive and develop a moderate degree of cardiac hypertrophy compared to wild type animals. In contrast, both ERα and ERβ have been invoked to play a relevant role in regulating vascular tone by increasing local NO release via genomic and nongenomic mechanisms (Rubanyi et al. 1997; Chambliss et al. 2000, 2002; Chen et al. 1999). Vascular remodeling has been studied extensively in ERα and ERβ knockout mice, until it finally became clear that ERα is the relevant estrogen receptor isoform to attenuate estrogen dependent neointima proliferation following carotid balloon injury in mice (Iafrati et al. 1997; Daras et al. 1999a,b; Pare et al. 2002). Importantly, ERα and ERβ effects are not limited to the vasculature but extend to cardiac muscle as well, since deletion of ERβ aggravates cardiac hypertrophy in mice following aortic banding (Skavkahl et al. 2004). Conversely, deletion of ERβ resulted in increased mortality and aggravation of heart failure in mice with experimental myocardial infarction (Pelzer et al. 2005b). Taken together, these observations suggest that ERα and ERβ play redundant, specific, and eventually also antagonistic roles in the cardiovascular system. Equally important, these studies suggest that the role of estrogens in heart disease cannot be understood without addressing the specific role and function of ERα and ERβ in different cardiovascular tissues and model systems of human heart disease.

Phenotypical differences in mice lacking either ERα or ERβ have stimulated research into isotype selective ERα and ERβ agonists, since such compounds might exert more specific and eventually improved pharmacological profiles compared to nonselective ER agonists such as 17β-estradiol, which binds and activates both ER isoforms with very similar affinity and potency (Harris et al. 2002). In addition, potent ERα and ERβ selective ER agonists would provide an independent and complementary strategy to differentiate the role of ERα and ERβ in animal models of human heart disease without altering the endogenous expression of their cognate receptors. Isotype selective activation of ERα and ERβ, which was first achieved using nonsteroidal ligands, is meanwhile possible also with newly synthesized steroidal ERα and ERβ agonists such as 16α-lactone-estradiol and 8β-vinyl-estradiol (Sun et al. 2003; Hillisch et al. 2004; Hegele-Hartung et al. 2004). Crystallographic analyses of the three-dimensional structure of the ERα and the ERβ ligand-binding pocket, revealed slightly more space below the steroidal D-ring of ERα, whereas the ERβ LBD left more space above the B- and C-ring. These structural differences allowed for a protein structure-based design of isotype selective ER agonists that resulted in potent and highly selective ERα and ERβ agonists that transactivate their specific target receptor with 190-fold (8β-VE2) or 265-fold (16α-LE2) higher potency (Elger and Fritzemeier 2004). The pharmacological properties of 16α-lactone-estradiol and 8β-vinyl-estradiol suggest that these compounds might serve as a novel tool to dissect the role of ERα and ERβ in the cardiovascular system without genetic manipulation of endogenous ER expression levels.

2 Studies of Isotype Selective ER Agonists in Spontaneously Hypertensive Rats

Spontaneously hypertensive rats (SHR) have been employed in numerous experimental studies of hypertension because this animal model mimics several characteristic features of human hypertensive heart disease, including vascular dysfunction and cardiac hypertrophy (Pelzer et al. 2002; Reckelhoff and Fortepiani 2004). Cardiac hypertrophy in young SHRs also responds well to sex hormone treatment, including nonselective ER agonists such as 17β-estradiol (Wassmann et al. 2001). Previous studies, which could not address the specific role of ERα and ERβ in cardiovascular pathology of SHR at the time, indicated that simultaneous activation of ERα and ERβ attenuates cardiac hypertrophy and improved impaired endothelium-dependent vascular relaxation in estrogen-depleted female SHRs. To clarify the relative importance of ERα and ERβ for cardiovascular morphology, function, and gene expression, ovariectomized SHRs were subjected to long-term treatment with 17β-estradiol or 16α-LE2. In line with the pharmacological profile of 16α-LE2 and as expected from uterus atrophy in ERKO but not in BERKO mice, 16α-LE2 attenuated uterus atrophy, (Fig. 1) (Emmen and Korach 2003). Similar results that were obtained for body weight measurements indicate that ligand-dependent activation of ERα is required to attenuate the adipose phenotype of estrogen-depleted SHRs (Fig. 1). Invasive hemodynamic analysis did not reveal a significant blood pressure lowering effect of 17β-estradiol and 16α-LE2 (Fig. 2). Hypertension increases cardiac afterload and triggers the activation of several cardiac signal transduction cascades that play an important role in the development of cardiac hypertrophy (Sussman et al. 2002). Cardiac hypertrophy by itself is an independent risk factor and predictor of cardiovascular mortality in human heart disease (Levy et al. 1990). Interestingly, selective activation of ERα by 16α-LE2 attenuated cardiac hypertrophy, although 16α-LE2 conferred no blood pressure lowering effect (Fig. 1). Accordingly, activation of ERα contents protective effects on cardiac mass. However, the mechanisms might at least in part be specific for individual isotype-selective ligands. In particular, inhibition of cardiac hypertrophy by 16α-LE2 is blood pressure-independent and was been linked to increased cardiac ANP expression. This hypothesis is supported by independent experiments that not only showed a causal role of ANP to attenuate cardiac mass, but also upregulation of ANP expression via activation of ERα in vitro (Pelzer et al. 2005a; Babiker et al. 2004).

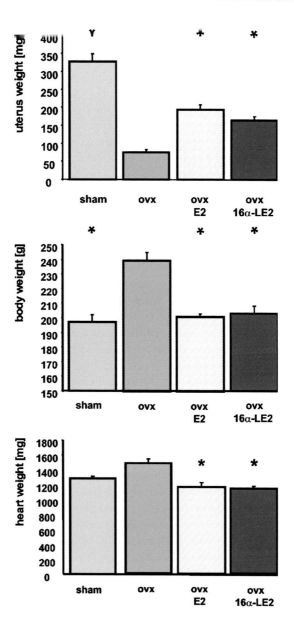

Fig. 1. Effects of 17β-estradiol and of 16α-LE2 on uterus weight, body weight, and heart weight in female spontaneously hypertensive rats (SHRs). Uterus atrophy in ovariectomized SHRs is prevented by 17β-estradiol and the ERα agonist 16α-LE2. Gain of body weight in ovariectomized SHRs is prevented by treatment with 17β-estradiol and 16α-LE2. Cardiac hypertrophy in estrogen-depleted SHRs is attenuated by estradiol and by 16α-LE2. (n = 8–10 animals/group, $*p = 0.05$)

◄

Longstanding cardiac hypertrophy frequently leads to impaired left ventricular function and thus acts as an important co-factor for the development of heart failure in humans and in animal models of human heart disease (Conrad et al. 1995). Although SHRs at a young age do not display clinical, hemodynamic, or biochemical signs of heart failure, it is nevertheless interesting to note that the ERα agonist 16α-LE2 improved cardiac output as well as left ventricular stroke volume in estrogen-depleted SHRs (Fig. 3). Although not functionally required in the current experimental setting, it appears conceivable that similar effects would augment cardiac function in animal models of chronic heart failure. Cardiac output as the product of heart rate and stroke volume depends on the contractile properties of individual cardiac myocytes, which are regulated by intracellular calcium homeostasis and by the structural composition of individual sarcomeres (Tardiff et al. 2000; Perez et al. 1999). In rodents, cardiac contractility is also regulated by differential expression of two distinct myosin heavy chain isoforms. The αMHC isoform, which is expressed under physiological conditions in the rodent heart, possesses a high ATPase activity and has also been termed the fast motor. In contrast, the slower and energetically economic βMHC isoform becomes the predominant MHC isoform in hypertrophied and failing rat hearts (Miyata et al. 2000). Selective activation of ERα by 16α-LE2 or nonselective activation of ERα and ERβ by 17β-estradiol prevented the loss of cardiac αMHC expression that occurred in ovariectomized SHRs also attenuated cardiac hypertrophy to an extent similar to E2 and 16α-LE2 (Fig. 4). These results support the hypothesis that cardiac αMHC expression in SHRs is a specific function of ERα, although both ER isoforms are robustly expressed in the

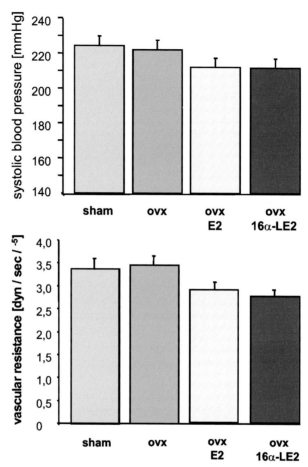

Fig. 2. Systolic blood pressure and systemic vascular resistance in female SHRs are not attected by treatment with estradiol or 16α-LE2. (n = 8–10 animals/group, *p = 0.05)

heart and the aorta of SHRs. These findings also indicate that 16α-LE2 increase cardiac output by an upregulation of cardiac αMHC expression (16α-LE2) and not by lowering blood pressure and vascular resistance.

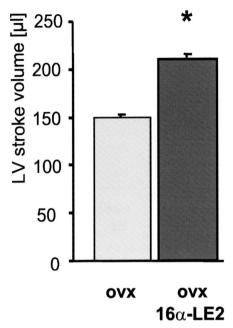

Fig. 3. Left ventricular stroke volume was higher in estrogen-depleted female SHRs treated with 16α-LE2 and 8β-VE2 compared to placebo (*ovx*). ($n = 8–10$ animals/group, $*p < 0.05$)

3 Studies of Isotype Selective ER Agonists in Animal Models of Aldosterone-Salt-Induced Cardiovascular Injury

Mineralocorticoids such as aldosterone act via ligand-dependent activation of the mineralocorticoid receptor (MR) and play an important physiological role in fluid and electrolyte homeostasis. In contrast to their physiological function, disproportionate elevation of serum aldosterone levels can result in a complex pattern of cardiovascular damage such as hypertension, cardiac hypertrophy, vascular inflammation, as well as vascular and cardiac fibrosis. These observations were initially reported in rodents receiving chronic aldosterone infusion and a high

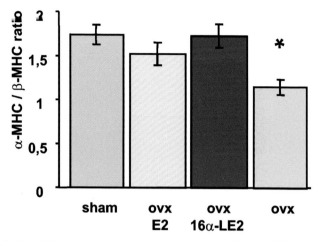

Fig. 4. The shift of cardiac isomyosins toward a predominant βMHC expression in ovariectomized SHRs is prevented by 17β-estradiol and the ERα agonist 16α-LE2 (n = 8–10 animals/group, $*p < 0.05$)

salt diet (Rocha et al. 2002). These findings were recently extended to the human situation because a relevant proportion of hypertensive individuals also exhibit a disproportionate elevation of aldosterone serum levels, which might play a causative role in salt-sensitive hypertension (Mulatero et al. 2004; Vasan et al. 2004). Excess mineralocorticoid activity also plays an important and unfavorable role in the development of chronic heart failure because disproportionate activation of the renin–angiotensin–aldosterone system promotes cardiac remodeling and fluid retention and hence aggravates the clinical symptoms of chronic heart failure (Swedberg et al. 1990). Accordingly, MR antagonists such as spironolactone, which was initially developed as a diuretic with potassium-sparing properties, have been successfully evaluated in the treatment of chronic heart failure (Pitt et al. 1999, 2003). Although mineralocorticoid and estrogen receptors are able to play important roles in cardiovascular function, very little is known on the functional interaction of these structurally related receptor systems. Therefore, and because estrogens and mineralocorticoids can confer very different and in part antagonistic effects in cardiovascular disease, one might speculate that activation of ERα and/or ERβ by isotype selective ER agonists might attenuate aldosterone-induced cardiovascular injury. To test this hypothesis, we analyzed cardiovascular function, morphology, and gene expression in female rats displaying disproportionate elevation of aldosterone serum levels due to chronic aldosterone infusion combined with a high-salt diet. The AST-rat model (aldosterone salt treatment) was developed by Brilla and co-workers in 1992 and exhibits systemic hypertension, cardiac hypertrophy, as well as vascular and cardiac fibrosis (Brilla and Weber 1992). AST rats were treated with placebo, estradiol, 16α-LE2, or 8β-VE2 for 2 months before a complete set of hemodynamic, morphometric, and molecular studies were performed. Cardiac hypertrophy in AST rats was significantly attenuated by estradiol subtype selective ER agonists (Fig. 5). The protective function of estrogens was not limited to the myocardium but also extended to the vascular bed because selective or nonselective activation of ERα and/or ERβ significantly attenuated excessive perivascular collagen accumulation in aldosterone–salt-treated rats (Fig. 6). Perivascular fibrosis, which is a very typical feature of excess MR activity or disproportionate MR activation by elevated aldosterone serum levels, is mediated by local infiltration of inflammatory cells and depends critically on increased expression of the integrin osteopontin (Young and Funder 2002). The latter hypothesis is supported by observations in osteopontin-null mice, which in contrast to WT mice do not develop perivascular fibrosis in response to aldosterone–salt treatment (Sam et al. 2004). Within this context, it is interesting to note that the osteopontin promoter has previously been shown to respond to estrogens in noncardiovascular cell types (Vanacker et al. 1999). Thus, the protective function of estrogens on perivascular fibrosis of AST rats could result from E2-dependent repression of osteopontin expression via activation of ERα and/or ERβ. As shown in Fig. 6, perivascular osteopontin levels were remarkably low in AST rats treated with 16α-LE2, 8β-VE2, or 17β-estradiol. These findings suggest that both ER subtypes attenuate vascular remodeling by lowering the local expression of the pro-inflammatory cytokine osteopontin. The mechanisms by which ERα and ERβ exert their protective effects under condition of excess MR activity are at least in part different from MR antagonists such as spironolactone and most likely do not interfere directly

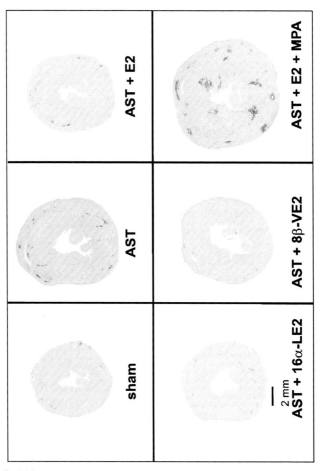

Fig. 5. Aldosterone–salt treatment (*AST*) for 8 weeks resulted in cardiac hypertrophy in ovariectomized, normotensive Wistar rats. 17β-estradiol, the ERα agonist 16α-LE2, the ERβ agonist 8β-VE2 attenuated cardiac mass in AST rats. MPA aggravated cardiac hypertrophy in E2 substituted rats. Representative photomicrographs of cardiac cross sections (picro-Sirius red stain)

Cardiovascular Effects of ERα and ERβ agonists

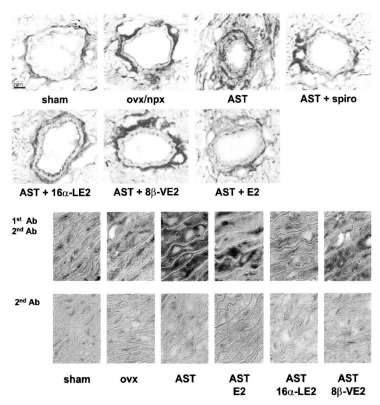

Fig. 6. Vascular remodeling, perivascular fibrosis, and vascular osteopontin expression in control rats (sham, *ovx/npx*) and AST rats, treated with placebo (*AST*), 17β-estradiol (AST + E2), 16α-LE2 (AST + 16α-LE2), 8β-VE2 (AST + 8β-VE2), or spironolactone (AST + spiro). Vascular remodeling and osteopontin accumulation are significantly attenuated by 17β-estradiol, 16α-LE2, and 8β-VE2; spironolactone conferred similar effects

with MR activation because the aldosterone-dependent suppression of angiotensin-II serum levels was not alleviated by 16α-LE2, 8β-VE2, or 17β-estradiol, whereas spironolactone treatment of AST rats resulted in higher AII serum levels (Fig. 7). Thus, it appears more likely that isotype selective or nonselective ER agonists interfere with downstream signal transduction pathways that become activated upon disproportionate MR activation. Initial insight into these mechanisms might come from studies that attempt to identify genes that are regulated by MR and/or ER activation in AST rats. A proteome-wide screen for cardiac proteins that are differentially regulated via MR and ER activation revealed a considerable number of proteins that were regulated in opposite directions by aldosterone and estrogens in AST rats. Interestingly, the expression of most of these peptides was downregulated with cardiac hypertrophy upon aldosterone–salt treatment, and estrogen treatment prevented the repression of a relevant proportion of these proteins, as shown in Fig. 8. Identification and functional characterization of these differentially expressed proteins is forthcoming and may provide further insight into the mechanisms by which isotype-selective and nonselective agonists protect the heart against excess MR activation.

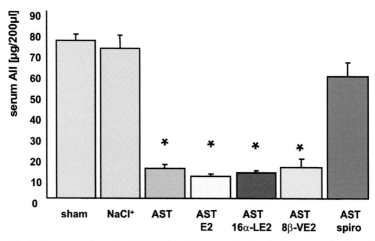

Fig. 7. Serum angiotensin-II (*AII*) levels in control animals (sham, *ovx/npx*) and AST rats upon long-term administration of placebo (*AST*), 17β-estradiol (AST + E2), 16α-LE2 (AST + 16α-LE2), 8β-VE2 (AST + 8β-VE2), or spironolactone (AST + spiro). Suppression of serum AII levels in AST rats was not affected by estrogen but alleviated by spironolactone. ($n = 8$–10 animals/group, * = $p < 0.05$)

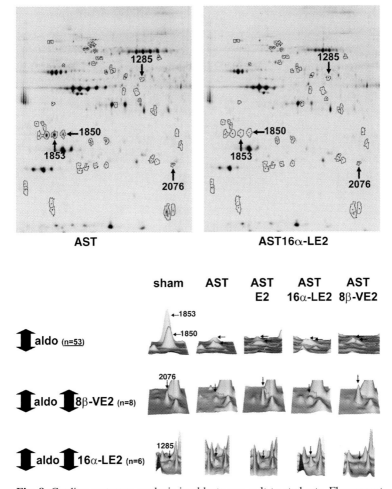

Fig. 8. Cardiac proteome analysis in aldosterone–salt-treated rats. Fluorescent 2DPAGE analysis revealed a total of 53 protein spots that showed differential expression between control (sham) and AST rats. Co-treatment of AST rats with 8β-VE2 blocked the differential expression of eight specific peptides, whereas the ERα agonist 16α-LE2 affected the expression of six specific peptides

4 Summary and Perspectives

Isotype selective agonists for ERα and ERβ such as 16α-LE2 and 8β-VE2 confer redundant as well as specific effects on cardiovascular function and gene expression in animal models of human cardiovascular disease. The observation that ligand-dependent activation of ERα and ERβ results in similar as well as distinct phenotypical changes in SHRs and aldosterone–salt-treated rats indicates that pharmacological activation of specific ER subtypes provides an alternative approach to transgenic mouse models to gain further insight into the specific function of ERα and ERβ in the cardiovascular system. Ultimately, isotype-selective ER ligands might also offer more specific and improved pharmacological profiles to optimize existing strategies and to develop novel treatment strategies.

References

Babiker FA, De Windt LJ, van Eickels M, Thijssen V, Bronsaer RJ, Grohe C, van Bilsen M, Doevendans PA (2004) 17beta-estradiol antagonizes cardiomyocyte hypertrophy by autocrine/paracrine stimulation of a guanylyl cyclase A receptor-cyclic guanosine monophosphate-dependent protein kinase pathway. Circulation 109:269–276

Barrett-Connor E, Bush TL (1991) Estrogen and coronary heart disease in women [see comments]. JAMA 265:1861–1867

Brilla CG, Weber KT (1992) Reactive and reparative myocardial fibrosis in arterial hypertension in the rat. Cardiovasc Res 26:671–677

Chambliss KL, Yuhanna IS, Anderson RG, Mendelsohn ME, Shaul PW (2002) ERbeta has nongenomic action in caveolae. Mol Endocrinol 16:938–946

Chen Z, Yuhanna IS, Galcheva-Gargova Z, Karas RH, Mendelsohn ME, Shaul PW (1999) Estrogen receptor alpha mediates the nongenomic activation of endothelial nitric oxide synthase by estrogen. J Clin Invest 103:401–406

Conrad CH, Brooks WW, Hayes JA, Sen S, Robinson KG, Bing OH (1995) Myocardial fibrosis and stiffness with hypertrophy and heart failure in the spontaneously hypertensive rat. Circulation 91:161–170

Emmen JM, Korach KS (2003) Estrogen receptor knockout mice: phenotypes in the female reproductive tract. Gynecol Endocrinol 17:169–176

Grady D, Herrington D, Bittner V, Blumenthal R, Davidson M, Hlatky M, Hsia J, Hulley S, Herd A, Khan S, Newby LK, Waters D, Vittinghoff E, Wenger N (2002) Cardiovascular disease outcomes during 6.8 years of hormone therapy: Heart and Estrogen/progestin Replacement Study follow-up (HERS II). JAMA 288:49–57

Green S, Walter P, Greene G, Krust A, Goffin C, Jensen E, Scrace G, Waterfield M, Chambon P (1986) Cloning of the human oestrogen receptor cDNA. J Steroid Biochem 24:77–83

Grodstein F, Stampfer M (1995) The epidemiology of coronary heart disease and estrogen replacement in postmenopausal women. Prog Cardiovasc Dis 38:199–210

Grohe C, Kahlert S, Lobbert K, Stimpel M, Karas RH, Vetter H, Neyses L (1997) Cardiac myocytes and fibroblasts contain functional estrogen receptors. FEBS Lett 416:107–112

Harris HA, Katzenellenbogen JA, Katzenellenbogen BS (2002) Characterization of the biological roles of the estrogen receptors ERalpha and ERbeta, in estrogen target tissues in vivo through the use of an ERalpha-selective ligand. Endocrinology 143:4172–4177

Hegele-Hartung C, Siebel P, Peters O, Kosemund D, Muller G, Hillisch A, Walter A, Kraetzschmar J, Fritzemeier KH (2004) Impact of isotype-selective estrogen receptor agonists on ovarian function. Proc Natl Acad Sci USA 101:5129–5134

Herrington DM, Howard TD (2003) From presumed benefit to potential harm—hormone therapy and heart disease. N Engl J Med 349:519–521

Hillisch A, Peters O, Kosemund D, Muller G, Walter A, Schneider B, Reddersen G, Elger W, Fritzemeier KH (2004) Dissecting physiological roles of estrogen receptor alpha and beta with potent selective ligands from structure-based design. Mol Endocrinol 18:1599–1609

Hulley S, Grady D, Bush T, Furberg C, Herrington D, Riggs B, Vittinghoff E (1998) Randomized trial of estrogen plus progestin for secondary prevention of coronary heart disease in postmenopausal women. Heart and Estrogen/progestin Replacement Study (HERS) Research Group. JAMA 280:605–613

Iafrati MD, Karas RH, Aronovitz M, Kim S, Sullivan TR Jr, Lubahn DB, O'Donnell TF Jr, Korach KS, Mendelsohn ME (1997) Estrogen inhibits the vascular injury response in estrogen receptor alpha-deficient mice. Nat Med 3:545–548

Karas RH, Schulten H, Pare G, Aronovitz MJ, Ohlsson C, Gustafsson JA, Mendelsohn ME (2001a) Effects of estrogen on the vascular injury response in estrogen receptor alpha, beta (double) knockout mice. Circ Res 89:534–539

Karas RH, van Eickels M, Lydon JP, Roddy S, Kwoun M, Aronovitz M, Baur WE, Conneely O, O'Malley BW, Mendelsohn ME (2001b) A complex role for the progesterone receptor in the response to vascular injury. J Clin Invest 108:611–618

Krege JH, Hodgin JB, Couse JF, Enmark E, Warner M, Mahler JF, Sar M, Korach KS, Gustafsson JA, Smithies O (1998) Generation and reproductive phenotypes of mice lacking estrogen receptor beta. Proc Natl Acad Sci USA 95:15677–15682

Kuiper GG, Enmark E, Pelto-Huikko M, Nilsson S, Gustafsson JA (1996) Cloning of a novel receptor expressed in rat prostate and ovary. Proc Natl Acad Sci USA 93:5925–5930

Kuiper GG, Carlsson B, Grandien K, Enmark E, Haggblad J, Nilsson S, Gustafsson JA (1997) Comparison of the ligand binding specificity and transcript tissue distribution of estrogen receptors alpha and beta. Endocrinology 138:863–870

Levy D, Garrison RJ, Savage DD, Kannel WB, Castelli WP (1990) Prognostic implications of echocardiographically determined left ventricular mass in the Framingham Heart Study. N Engl J Med 322:1561–1566

Lindberg MK, Moverare S, Skrtic S, Gao H, Dahlman-Wright K, Gustafsson JA, Ohlsson C (2003) Estrogen receptor (ER)-beta reduces ERalpha-regulated gene transcription, supporting a "ying yang" relationship between ERalpha and ERbeta in mice. Mol Endocrinol 17:203–208

Lubahn DB, Moyer JS, Golding TS, Couse JF, Korach KS, Smithies O (1993) Alteration of reproductive function but not prenatal sexual development after insertional disruption of the mouse estrogen receptor gene. Proc Natl Acad Sci USA 90:11162–11166

Mulatero P, Stowasser M, Loh KC, Fardella CE, Gordon RD, Mosso L, Gomez-Sanchez CE, Veglio F, Young WF Jr (2004) Increased diagnosis of primary aldosteronism, including surgically correctable forms, in centers from five continents. J Clin Endocrinol Metab 89:1045–1050

Miyata S, Minobe W, Bristow MR, Leinwand LA (2000) Myosin heavy chain isoform expression in the failing and nonfailing human heart. Circ Res 86:386–390

Pare G, Krust A, Karas RH, Dupont S, Aronovitz M, Chambon P, Mendelsohn ME (2002) Estrogen receptor-alpha mediates the protective effects of estrogen against vascular injury. Circ Res 90:1087–1092

Pelzer T, de Jager T, Muck J, Stimpel M, Neyses L (2002) Oestrogen action on the myocardium in vivo: specific and permissive for angiotensin-converting enzyme inhibition. J Hypertens 20:1001–1006

Pelzer T, Jazbutyte V, Hu K, Segerer S, Nahrendorf M, Nordbeck P, Bonz AW, Muck J, Fritzemeier KH, Hegele-Hartung C, Ertl G, Neyses L (2005a) The estrogen receptor-alpha agonist 16alpha-LE2 inhibits cardiac hypertrophy and improves hemodynamic function in estrogen-deficient spontaneously hypertensive rats. Cardiovasc Res 67:604–612

Pelzer T, Loza PA, Hu K, Bayer B, Dienesch C, Calvillo L, Couse JF, Korach KS, Neyses L, Ertl G (2005b) Increased mortality and aggravation of heart failure in estrogen receptor-beta knockout mice after myocardial infarction. Circulation 111:1492–1498

Perez NG, Hashimoto K, McCune S, Altschuld RA, Marban E (1999) Origin of contractile dysfunction in heart failure: calcium cycling versus myofilaments. Circulation 99:1077–1083

Pitt B, Zannad F, Remme WJ, Cody R, Castaigne A, Perez A, Palensky J, Wittes J (1999) The effect of spironolactone on morbidity and mortality in patients with severe heart failure. Randomized Aldactone Evaluation Study Investigators. N Engl J Med 341:709–717

Pitt B, Remme W, Zannad F, Neaton J, Martinez F, Roniker B, Bittman R, Hurley S, Kleiman J, Gatlin M (2003) Eplerenone, a selective aldosterone blocker, in patients with left ventricular dysfunction after myocardial infarction. N Engl J Med 348:1309–1321

Reckelhoff JF, Fortepiani LA (2004) Novel mechanisms responsible for postmenopausal hypertension. Hypertension 43:918–923

Rossouw JE, Anderson GL, Prentice RL, LaCroix AZ, Kooperberg C, Stefanick ML, Jackson RD, Beresford SA, Howard BV, Johnson KC, Kotchen JM, Ockene J (2002) Risks and benefits of estrogen plus progestin in healthy postmenopausal women: principal results from the Women's Health Initiative randomized controlled trial. JAMA 288:321–333

Rubanyi GM, Freay AD, Kauser K, Sukovich D, Burton G, Lubahn DB, Couse JF, Curtis SW, Korach KS (1997) Vascular estrogen receptors and endothelium-derived nitric oxide production in the mouse aorta. Gender difference and effect of estrogen receptor gene disruption. J Clin Invest 99:2429–2437

Sam F, Xie Z, Ooi H, Kerstetter DL, Colucci WS, Singh M, Singh K (2004) Mice lacking osteopontin exhibit increased left ventricular dilation and reduced fibrosis after aldosterone infusion. Am J Hypertens 17:188–193

Skavdahl M, Steenbergen C, Clark J, Myers P, Demianenko T, Mao L, Rockman HA, Korach KS, Murphy E (2004) Beta estrogen receptor mediates male-female differences in the development of pressure overload hypertrophy. Am J Physiol Heart Circ Physiol 288:H4699–H476

Stampfer MJ, Colditz GA, Willett WC, Manson JE, Rosner B, Speizer FE, Hennekens CH (1991) Postmenopausal estrogen therapy and cardiovascular disease. Ten-year follow-up from the nurses' health study. N Engl J Med 325:756–762

Sun J, Baudry J, Katzenellenbogen JA, Katzenellenbogen BS (2003) Molecular basis for the subtype discrimination of the estrogen receptor-beta-selective ligand, diarylpropionitrile. Mol Endocrinol 17:247–258

Sussman MA, McCulloch A, Borg TK (2002) Dance band on the Titanic: biomechanical signaling in cardiac hypertrophy. Circ Res 91:888–898

Swedberg K, Eneroth P, Kjekshus J, Wilhelmsen L (1990) Hormones regulating cardiovascular function in patients with severe congestive heart failure and their relation to mortality. CONSENSUS Trial Study Group. Circulation 82:1730–1736

Tardiff JC, Hewett TE, Factor SM, Vikstrom KL, Robbins J, Leinwand LA (2000) Expression of the beta (slow)-isoform of MHC in the adult mouse heart causes dominant-negative functional effects. Am J Physiol Heart Circ Physiol 278:H412–H419

Vanacker JM, Pettersson K, Gustafsson JA, Laudet V (1999) Transcriptional targets shared by estrogen receptor- related receptors (ERRs) and estrogen receptor (ER) alpha, but not by ERbeta. EMBO J 18:4270–4279

Vasan RS, Evans JC, Benjamin EJ, Levy D, Larson MG, Sundstrom J, Murabito JM, Sam F, Colucci WS, Wilson PW (2004) Relations of serum aldosterone to cardiac structure: gender-related differences in the Framingham Heart Study. Hypertension 43:957–962

Wassmann S, Baumer AT, Strehlow K, van Eickels M, Grohe C, Ahlbory K, Rosen R, Bohm M, Nickenig G (2001) Endothelial dysfunction and oxidative stress during estrogen deficiency in spontaneously hypertensive rats. Circulation 103:435–441

Young MJ, Funder JW (2002) Mineralocorticoid receptors and pathophysiological roles for aldosterone in the cardiovascular system. J Hypertens 20:1465–1468

Pathogenesis and Therapy of Rheumatoid Arthritis

R.O. Williams(✉)

Kennedy Institute of Rheumatology Division, Imperial College London, 1 Aspenlea Road, W6 8LH London, UK
email: *richard.o.williams@imperial.ac.uk*

1	Introduction	108
2	Etiopathogenesis of RA	108
2.1	Genes Versus the Environment	108
2.2	Evidence of a Role for T Lymphocytes in RA	109
2.3	Cytokine Expression in RA	110
3	The Role of TNFα: Lessons from Animal Models	113
3.1	hTNF Transgenic Mice	113
3.2	TNFα Blockade in Collagen-Induced Arthritis	114
3.3	TNFα Blockade in Human RA	115
4	What Has Anti-TNFα Therapy Taught Us About the Pathogenesis of RA?	117
4.1	The Role of Angiogenesis	117
4.2	TNFα Drives Inflammatory Cell Recruitment	118
4.3	Evidence of Aberrant T Cell Signaling in RA	119
4.4	Is There a Failure in Regulatory T Cell Activity in RA?	120
5	Are Female Sex Hormones Protective in RA?	121
5.1	Evidence from Epidemiological Studies	121
5.2	Evidence from Animal Models	122
5.3	Future Perspectives: Selective ER Agonists	123
References		123

Abstract. Rheumatoid arthritis is a chronic disabling disease affecting at least 1% of the population on a worldwide basis. Research aimed at understanding the pathogenesis of this disease led to the identification of TNFα as a major pro-inflammatory cytokine expressed in the inflamed joints of patients with rheumatoid arthritis. Subsequently, in vitro studies provided evidence to suggest that TNFα played an important role in driving the expression of additional pro-inflammatory cytokines, such as IL-1, GM-CSF, IL-6, and IL-8, in synovial cell cultures. Another important finding that confirmed the pathological significance of TNFα was that mice genetically engineered to overexpress TNFα spontaneously developed arthritis. Subsequently, the therapeutic effect TNFα blockade was tested in animal models prior to clinical trials in human patients, which provided unequivocal verification of the validity of TNFα as a therapeutic target. Anti-TNFα therapy is now accepted as a fully-validated treatment modality for rheumatoid arthritis.

1 Introduction

Rheumatoid arthritis (RA) is a chronic inflammatory disease of unknown etiology that is often progressive in nature, resulting in pain, stiffness, and swelling of joints. In late stages, deformity and ankylosis develop. The worldwide prevalence of RA is around 1%, with a female-to-male ratio of approximately 3 to 1. Although RA can strike at any age, the age of onset of RA is typically between 25 and 50. RA is one of the most common causes of disability in the Western world and pathological features of the disease include inflammatory erosive synovitis that causes destruction of cartilage, bone, and soft tissues and this can result in deformity and loss of joint function. RA is usually classified as a non-organ-specific autoimmune disease because, in addition to joint involvement, extraarticular features, such as subcutaneous nodules, vasculitis, and pulmonary fibrosis, are not uncommon, especially in the more severe cases.

2 Etiopathogenesis of RA

2.1 Genes Versus the Environment

It has been known for many years that genetic factors are important in determining susceptibility to RA, and the role of HLA DRB1 alleles as a major risk factor for the disease is well established. Thus, associations between different HLA DRB1 alleles have been demonstrated in a number of ethnic groups around the world (Silman and Pearson 2002). It should also be borne in mind, however, that data from twin studies (in which concordance is only about 15% (Silman et al. 1993)), have suggested that the maximum potential genetic contribution to disease susceptibility is approximately 50%–60%. Furthermore, only around 50% of the genetic contribution to RA can be explained by HLA genes, and therefore non-MHC genes should not be overlooked in the search for causes of the disease.

A large number of environmental, infectious and hormonal factors have been proposed as contributors to susceptibility to RA. For example, there is evidence that female sex hormones are protective, based on the observations that the use of the oral contraceptive pill and pregnancy are both associated with a reduced risk of developing RA. In contrast, the postpartum period is associated with an increased risk of developing RA (Silman and Pearson 2002). Although infection has been proposed as a potential trigger for RA, there is no firm epidemiological data to support this hypothesis at present. There is, however, mounting evidence to suggest a link between cigarette smoking and the development of RA (Albano et al. 2001; Harrison et al. 2001; Hutchinson et al. 2001; Karlson et al. 1999). A similar association has also been described for Crohn disease.

2.2 Evidence of a Role for T Lymphocytes in RA

It has been suspected for more than a quarter of a century that $CD4^+$ T cells play an important role in the pathogenesis of RA (Janossy et al. 1981). This suspicion was based principally on the presence of T cells in RA synovium and on the well-established HLA class II association with RA (Todd et al. 1985) because the only known function of MHC class II molecules is to present antigens to $CD4^+$ T cells. Further evidence to support the role of T cells was provided by experiments with animal models in which a variety of different antibodies, including anti-CD4 (Ranges et al. 1985), anti-TCR (Goldschmidt and Holmdahl 1991; Yoshino et al. 1991), anti-IL-2R (Banerjee et al. 1988) and anti-MHC class II (Cooper et al. 1988) were found to inhibit disease development.

These studies were influential in the establishment of clinical trials of depleting anti-CD4 mAb therapy in RA, but paradoxically, these trials were not on the whole successful, despite achieving a high rate of CD4$^+$ blood T cell depletion (Choy et al. 1998; Moreland et al. 1997b). Some nondepleting anti-CD4 mAbs have shown transient beneficial effects however (Choy et al. 1998), although the relatively low efficacy as well as the presence of side effects (e.g., vasculitis) have led to the discontinuation of this therapeutic strategy.

It has been shown in a number of human autoimmune diseases that the expression of HLA class II molecules is increased in diseased tissue compared to normal tissue, thereby providing the opportunity for presentation of antigens to CD4$^+$ T cells. Furthermore, in some disease states (e.g., Graves' autoimmune hyperthyroidism) it was shown that the increased expression of MHC class II molecules extended even to cells that do not normally express MHC class II molecules, including thyroid epithelial cells (Bottazzo et al. 1983; Hanafusa et al. 1983).

In RA, endothelial cells and fibroblasts have been shown to exhibit increased MHC class II molecule expression, and this may be interpreted as evidence of increased antigen-presenting cell (APC) function (Janossy et al. 1981; Klareskog et al. 1982). As cytokines (e.g., IFNγ) are the principal inducers of upregulated MHC class II expression, it was proposed that aberrant expression of cytokines led to increased APC activity, resulting in the presentation of self-antigens and the development of autoimmune disease (Bottazzo et al. 1983). Further evidence that increased APC function was important in organ-specific autoimmune disease was provided by the studies of Londei et al. in Grave's disease (Londei et al. 1984, 1985). These early studies provided the rationale for the cataloguing of cytokines in the joints of patients with RA in order to identify candidates with the potential to contribute to the pathogenesis of the disease.

2.3 Cytokine Expression in RA

A number of different assays have been used to identify cytokines in RA synovium, including Northern blotting, slot blotting, and in situ hybridization for cytokine mRNA. For the detection of cytokine protein, ELISA assays and immunohistochemistry have been employed using monoclonal and polyclonal antibodies. These studies have shown that the majority of known cytokines can be detected in active rheumatoid synovium (Table 1) (Buchan et al. 1988; Hopkins and Meager 1988; Malyak et al. 1993; Symons et al. 1988; Xu et al. 1989). However, one cytokine usually found to be absent from the joint is IL-4 (O'Garra and Murphy 1993), and this may explain the Th1-skewing that occurs in RA patients.

It is of interest that proinflammatory cytokines, including TNFα, IL-1α/β, IL-6, and GM-CSF were present in virtually all samples, irrespective of the stage of disease (Feldmann and Maini 2001; Feldmann et al. 1996). These proinflammatory cytokines are normally expressed in a very transient manner and it was hypothesized that proinflammatory cytokines were being continuously overproduced in the RA joint (Feldmann and Maini 2001; Feldmann et al. 1996). In order to validate this hypothesis, short-term cultures of synovial tissue were established and culture supernatants assayed for spontaneous cytokine production.

The cells in the cultures of dissociated rheumatoid synovium were found to produce relatively large quantities of multiple cytokines, cytokine inhibitors, matrix metalloproteinases, etc., in a manner comparable to that observed in situ. In a highly influential series of experiments, it was subsequently shown that inhibition of TNFα in synovial cell cultures led to marked downregulation of the expression of IL-1 and other pro-inflammatory cytokines (Alvaro-Garcia et al. 1991; Brennan et al. 1989, 1990; Butler et al. 1995). These observations pointed to the presence of a cytokine cascade in which TNFα was responsible for driving the production of a number of pro-inflammatory mediators.

It is clear that in RA joints there is abundant expression of pro-inflammatory cytokines but another possibility that has to be considered is that in RA there is a failure in the expression of anti-inflammatory mediators. However, this is unlikely to be the case because multiple anti-inflammatory mediators, including IL-10 (Cohen et al. 1995), IL-11 (Hermann et al. 1998), soluble TNF receptor (Brennan et al. 1995; Cope et al. 1992), IL-1 receptor antagonist (Deleuran et al. 1992), and TGFβ (Chu et al. 1991) have been found to be upregulated in RA synovium. Furthermore, neutralization of these anti-inflammatory mediators in culture supernatants leads to an increase in the production of IL-1 and TNFα (Cohen et al. 1995; Hermann et al. 1998), indicating that that they are fully functional. This led Feldmann and co-workers to propose that in RA there is an imbalance between pro- and anti-inflammatory

Table 1 Cytokines expressed in synovial tissue from patients with RA (adapted from Feldmann et al. 1996)

Cytokine	Expression	
	mRNA	Protein
Pro-inflammatory		
IL-1α, β	+	+
TNF	+	+
Lymphotoxin	+	±
IL-6	+	+
GM-CSF	+	+
M-CSF	+	+
LIF	+	+
Oncostatin M	+	+
IL-12	+	+
IL-15	+	+
IFNα, β	+	+
IFNγ	+	±
IL-17	+	+
IL-18	+	+
Immunoregulatory		
IL-2	+	±
IL-4	±	
IL-10	+	∓
IL-11	+	+
IL-13	+	+
TGFβ	+	+
Chemokins		
IL-8	+	+
Gro α	+	+
MIP-1	+	+
MCP-1	+	+
ENA-78	+	+
RANTES	+	+
Growth Factors		
FGF	+	+
PDGF	+	+
VEGF	+	+

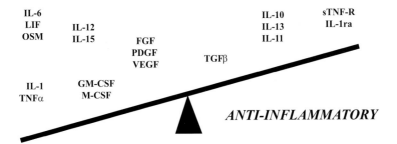

PRO-INFLAMMATORY

Fig. 1. Disequilibrium of pro- and anti-inflammatory cytokines and chemokines in RA synovium. (Feldmann et al. 2003)

factors, with pro-inflammatory cytokines dominating over those that play an anti-inflammatory role, and this leads to chronic inflammation (Fig. 1).

3 The Role of TNFα: Lessons from Animal Models

3.1 hTNF Transgenic Mice

As discussed above, blockade of TNFα in synovial cell cultures leads to inhibition of IL-1 and other pro-inflammatory cytokines. Further evidence for an important role for TNFα was provided by the finding that mice overexpressing a human TNFα transgene, dysregulated by the replacement of the 3′ AU rich region with the 3′ untranslated region of the human β-globin gene spontaneously developed arthritis (Keffer et al. 1991). Treatment with anti-human anti-TNF mAb blocked the development of disease. Histological examination of the joints of hTNFα transgenic mice revealed that the arthritis bore a number of important similarities to human RA and was highly erosive in nature, with subchondral bone being a particularly prominent feature (Fig. 2).

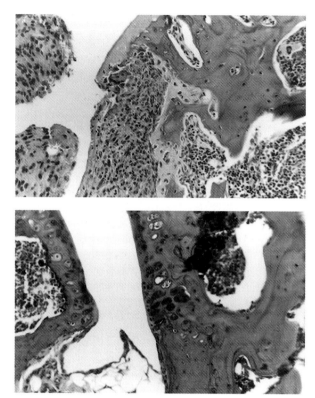

Fig. 2. Joint damage in hTNFα-transgenic mice. *Top*: erosive changes in the cartilage-bone-pannus region of a proximal interphalangeal joint from a hTNFα-transgenic mouse with arthritis. Note the focal erosion of subchondral bone. *Bottom*: normal joint from a nontransgenic littermate. H&E

3.2 TNFα Blockade in Collagen-Induced Arthritis

A number of studies have been conducted to assess the effect of TNFα blockade in collagen-induced arthritis (CIA), which is the most widely utilized animal model for human RA. These studies showed that treatment of mice with monoclonal or polyclonal anti-TNFα antibodies, or soluble TNF receptors, reduced the severity of arthritis when administered before the onset of clinical disease (Piguet et al. 1992; Thorbecke et al. 1992; Williams et al. 1992). Subsequently, we evaluated the effect of anti-TNFα treatment in mice with established arthritis (Williams et al. 1992). DBA/1 mice with CIA were given twice-weekly injections of TN3-19.12 (anti-TNFα mAb), L2 (isotype control), or PBS over a period of 14 days. The half-life of TN3-19.12 in mice had been previously estimated to be approximately 7 days (Sheehan et al. 1989). A dose-dependent reduction in the severity of arthritis following treatment with anti-TNFα mAb was observed (Fig. 3).

At the end of the treatment period, individual joints were graded according to the histopathological severity of arthritis. Anti-TNFα treatment was found to reduce the histological severity of arthritis and to protect joints from erosive changes (Fig. 4).

Soluble TNF receptors are understood to play an important physiological role in regulating the activity of TNFα, and it was subsequently shown that two soluble TNFR constructs were effective in established CIA. In the first study, a p75 TNFR-Fc fusion protein was found to reduce the severity of CIA whether given before or after the onset of the disease (Wooley et al. 1993). In another study, we showed that a p55 TNFR-Ig fusion protein was effective in reducing both the clinical severity of established CIA (Williams et al. 1995). Furthermore, when the joints were examined by histology, treatment with TNFR-Ig was found to have exerted a dose-dependent protective effect on joint erosion. These findings confirmed the importance of TNFα in CIA and provided the rationale for the testing of anti-TNFα antibody therapy in human RA.

3.3 TNFα Blockade in Human RA

The importance of TNFα in the pathogenesis of RA was finally confirmed in clinical trials in which intravenous administration of chimeric anti-TNFα mAb (infliximab, Remicade) caused clear reductions in the level of disease activity and radiographic progression of disease (Elliott et al. 1993, 1994a,b). Similar findings were subsequently reported for soluble TNF receptor-Fc fusion protein (etanercept, Enbrel) (Hasler et al. 1996; Moreland et al. 1997a; Weinblatt et al. 1999). In addition to confirming the importance of TNFα in RA, these clinical trials provided

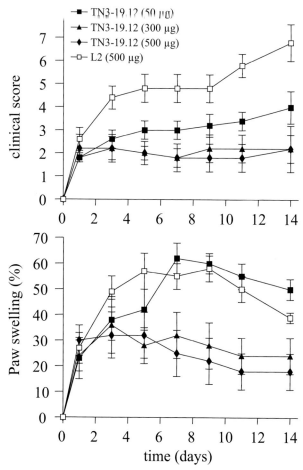

Fig. 3. Effect of anti-TNFα mAb (TN3-19.12) on clinical progression of established CIA. L2 is an isotype-matched control mAb. *Arrows* indicate time of injection. *Top*: clinical score, using a scoring system, was based on the following criteria: 0 = normal, 1 = slight swelling and/or erythema, 2 = pronounced edematous swelling, 3 = ankylosis. Each limb was graded, giving a maximum score of 12 per mouse. *Bottom*: paw-swelling, expressed as the percentage increment in paw-width relative to the paw-width before the onset of arthritis. (Modified from Williams et al. 1992)

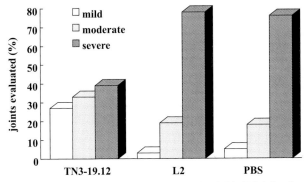

Fig. 4. Histopathological assessment of joints of arthritic DBA/1 mice treated with anti-TNFα. The scoring system was as follows. Mild, minimal synovitis, erosions limited to discrete foci, cartilage surface intact. Moderate, synovitis and erosions present but normal joint architecture intact. Severe, extensive erosions, joint architecture disrupted. (Data from Williams et al. 1992)

a unique opportunity to study the mechanism of benefit of TNFα blockade, and this has provided important information about the pathogenesis of the disease process, as discussed below.

4 What Has Anti-TNFα Therapy Taught Us About the Pathogenesis of RA?

4.1 The Role of Angiogenesis

Angiogenesis plays a major role in numerous pathological conditions, including joint inflammation, tumor growth and metastasis, diabetic retinopathy, and atherosclerosis. Evidence of a role for angiogenesis in the pathogenesis of RA is provided by a number of findings, including the observation that exuberant angiogenesis is usually seen in joint tissue from RA patients (Hirohata and Sakakibara 1999) and that blood vessel number in the synovium correlates positively with the level of hyperplasia, infiltration of mononuclear cells, and severity of joint tenderness (Rooney et al. 1988). Of course, such studies do not tell us whether angiogenesis arises as a consequence of inflammation or is a cause of the inflammation. However, one recent study in RA suggests a causative role for angiogenesis by demonstrating that there is an increase in angiogenesis prior to the early pathological changes in the synovium, including inflammatory cell infiltration and synoviocyte hyperplasia (Sivakumar et al. 2004).

A number of pro-angiogenic cytokines and growth factors have been identified in the joints of RA patients (Afuwape et al. 2002). Indeed, circulating levels of vascular endothelial growth factor (VEGF), which is often regarded as a master regulator of angiogenesis, are increased in RA compared to healthy individuals or patients with osteoarthritis (Ballara et al. 2001). Furthermore, levels of VEGF at presentation have been reported to show a positive correlation with the degree of deterioration of joints observed by radiography during the 1st year, indicating that circulating VEGF levels are predictive of future disease progression (Ballara et al. 2001).

Clinical trials of anti-TNFα mAb therapy in RA demonstrated a reduction in serum levels of VEGF after therapy. Furthermore, combined treatment with anti-TNFα mAb plus methotrexate in multiple infusions extended the period during which VEGF levels were reduced (Paleolog et al. 1998). From these findings it is suggested that at least one of the mechanisms by which TNFα blockade exerts its therapeutic effect is to reduce levels of VEGF expression, resulting in a subsequent reduction in angiogenesis and suppression of inflammation. This is consistent with the finding that inhibition of VEGF in established CIA reduces joint inflammation (Afuwape et al. 2002; Miotla et al. 2000; Sumariwalla et al. 2003).

4.2 TNFα Drives Inflammatory Cell Recruitment

One of the mechanisms underlying the potent anti-inflammatory effect of anti-TNFα mAb therapy in RA is that it reduces the influx of inflammatory cells to the joint (Taylor et al. 2000). This conclusion is based on a study in which a group of ten RA patients were given a single 10-mg/kg intravenous infusion of anti-TNFα mAb (infliximab) and the accumulation of autologous granulocytes (separated in vitro and labeled with ^{111}In) in the joint was analyzed using gamma-camera images before and after treatment. Synovial biopsies were also assessed at the same time points in order to identify infiltrating CD3$^+$ T cells, CD22$^+$ B cells, and CD68$^+$ macrophages by immunohistochemistry, and synovial tissue was studied for the presence of chemokines. Circulating levels of IL-8 and MCP-1 were also measured. TNFα blockade led to a significant reduction in the infiltration of ^{111}In-labeled granulocytes into arthritic joints and a significant reduction in infiltrating T cells, B cells, and macrophages. These reductions in numbers of infiltrating leukocytes were paralleled by reductions in chemokine expression and it was concluded that blockade of TNFα leads to reduced chemokine expression and reduced infiltration of inflammatory cells into the inflamed joint (Taylor et al. 2000).

4.3 Evidence of Aberrant T Cell Signaling in RA

RA is generally thought of as a disease of lymphocyte hyperresponsiveness that results in excessive pro-inflammatory cytokine expression. However, we now know that subpopulations of lymphocytes exist with the capacity to suppress inflammatory responses and another way of approaching the problem is to address the question of why, in susceptible individuals, do endogenous regulatory mechanisms fail to downregulate the inflammatory process. One possible explanation being explored by Cope et al. is that prolonged exposure of lymphocytes to high levels of TNFα within the joint leads to a state of T cell hyporesponsiveness to stimulation via the TCR and the uncoupling of proximal TCR signal transduction pathways. One possible consequence of this is a failure of T cell-driven immunoregulatory responses, prolonged inflammatory cell survival times, and increased effector responses (Cope 2003).

To validate the hypothesis that chronic exposure to TNFα downregulates T cell activity in RA, tetanus toxoid-specific T cell clones were pretreated with TNFα for up to 16 days prior to antigenic challenge. It was confirmed that antigen-specific proliferation was downregulated by TNFα in a dose- and time-dependent fashion, while responses to IL-2 and PHA were not significantly affected (Cope et al. 1994). It was also shown that pretreatment of T cells with TNFα resulted in significantly reduced production of IL-2, IL-10, IFNγ, TNFα, and lymphotoxin following anti-CD3 mAb stimulation, and reduced IL-2R alpha chain expression. Consistent with these findings was the observation that pretreatment of T cells with neutralizing anti-TNFα mAb enhanced proliferative responses, increased lymphokine production, and IL-2R alpha chain expression following mitogenic stimulation with anti-CD3 mAb.

Next, the immunomodulatory role played by TNFα on T cell activity was confirmed in vivo. This was achieved by analyzing cell-mediated immune responses in RA patients before and after treatment with TNF-blocking antibodies. TNFα blockade was found to normalize the diminished T cell proliferative responses both to mitogens and recall antigens in all patients tested. These in vivo findings confirm that chronic overproduction of TNFα does indeed blunt cell-mediated immune responses in vivo but it remains to be established whether this is responsible for suboptimal regulatory T cell activity (Cope et al. 1994).

4.4 Is There a Failure in Regulatory T Cell Activity in RA?

Studies with animal models clearly implicate a role for regulatory T cells in the prevention and control of autoimmunity (Powrie et al. 1994; Sakaguchi et al. 1995; Shevach 2000), although similar data in human autoimmune diseases is still lacking. In an interesting study by Ehrenstein et al., the question was addressed of whether regulatory T cell activity was normal or abnormal in patients with RA (Ehrenstein et al. 2004). It was shown that regulatory CD4$^+$CD25$^+$ T cells derived from RA patients exhibited the same anergic phenotype as regulatory T cells from normal individuals and were able to suppress the proliferative responses of CD4$^+$CD25$^-$ effector T cells in vitro. However, CD4$^+$CD25$^+$ T cells from RA patients were unable to suppress pro-inflammatory cytokine production by T cells stimulated with anti-CD3 or monocytes stimulated with LPS and, unlike CD4$^+$CD25$^+$ T cells from normal individuals, did not show evidence of infectious tolerance when cultured in the presence of effector CD4$^+$CD25$^-$ T cells. However, following treatment with anti-TNFα mAb there was an increase in the capacity of CD4$^+$CD25$^+$ regulatory T cells to downregulate cytokine production and to confer suppressive properties to effector T cells. Another important observation was that TNFα blockade resulted in increased numbers of circulating regulatory T cells in RA patients and this was paralleled by a reduction in the acute phase protein, C reactive protein.

This study suggests, firstly, that the activity of regulatory T cells is diminished in RA and, secondly, that blockade of TNFα causes both an increase in numbers and activity of $CD4^+CD25^+$ T cells (Ehrenstein et al. 2004). However, in another different study, it was shown that the percentage of $CD4^+CD25^+$ T cells in the synovial fluid of RA patients was significantly increased compared with RA peripheral blood (suggesting local accumulation of regulatory T cells at the site of disease activity), and that the percentage of circulating $CD4^+CD25^+$ T cells in RA patients was higher than that found in healthy individuals (van Amelsfort et al. 2004). Furthermore, the $CD4^+CD25^+$ regulatory T cells in RA were comparable both phenotypically and functionally to $CD4^+CD25^+$ regulatory T cells from healthy controls (van Amelsfort et al. 2004). Further research will clearly be required to establish the precise role of regulatory T cell populations in RA and whether it will be possible to expand or activate these populations through therapeutic intervention.

5 Are Female Sex Hormones Protective in RA?

5.1 Evidence from Epidemiological Studies

Evidence from a number of studies has suggested that oral contraceptives (usually containing a combination of synthetic derivatives of estrogen and progesterone) are protective in RA (Anonymous 1978; Brennan et al. 1997), although current contraceptive use appears to be more important than ex-contraceptive use. One recent study has also shown that oral contraceptive use does not significantly affect disease outcome in the long term, suggesting that contraceptives delay, rather than prevent, disease development (Drossaers-Bakker et al. 2002). Pregnancy has also been associated with a reduction in disease activity followed by disease flare during the postpartum period as well as an increased risk of developing RA, particularly after the first pregnancy (Da Silva and Spector 1992; Nelson and Ostensen 1997; Silman et al. 1992).

5.2 Evidence from Animal Models

Studies in the CIA model have clearly demonstrated disease remission during pregnancy, followed by exacerbation postpartum (Hirahara et al. 1986; Waites and Whyte 1987; Whyte and Waites 1987). Similar findings have been reported in adjuvant arthritis in rats and proteoglycan-induced arthritis in BALB/c mice (Buzas et al. 1993; Mathers and Russell 1990). Pregnancy-induced remission in CIA is associated with a reduction in circulating levels of all type II collagen-specific isotypes of IgG (Williams and Whyte 1989). In order to investigate possible reasons for the postpartum exacerbation of arthritis, we administered bromocriptine mice with CIA immediately postpartum and found that the drug suppressed the clinical exacerbation of disease. Thus, there was approximately a 50% reduction in severity of disease in bromocriptine-treated animals compared to untreated mice. The effect was thought to be due to suppression of the prolactin release that normally occurs postpartum. However, in a more comprehensive study by Mattsson et al., it was reported that the post-partum exacerbation in arthritis is more likely to be due to a fall in estrogen levels, rather than a surge in prolactin (Mattsson et al. 1991).

In another experiment, evidence was presented to suggest that estrogen, but not progesterone, may be the critical factor to explain the pregnancy-associated remission of CIA (Jansson and Holmdahl 1989). This finding led to the concept that estrogen (or synthetic derivatives of estrogen) would be therapeutically effective in arthritis and this was subsequently confirmed to be the case in both male and female mice (Jansson and Holmdahl 1992; Jansson et al. 1990, 1994). Conversely, blockade of the estrogen receptor (ER) using ICI 182,780 was found to exacerbate CIA (Jansson and Holmdahl 2001).

Analysis of the anti-collagen immune response in 17β-estradiol-treated mice revealed that T cell production of IFNγ in lymph node cells was decreased, and significant decreases were also observed in levels of IL-10 and GM-CSF produced by lymph node cells from estradiol-treated mice. Although the total IgG anti-CII response was only minimally affected by estrogen treatment, a significant reduction in the levels of IgG2a anti-CII Abs and an increase in the levels of IgG1 anti-CII Abs were observed in estradiol-treated mice.

5.3 Future Perspectives: Selective ER Agonists

The ER exists in two main forms, ERα and ERβ, which have distinct tissue expression profiles. Furthermore, gene-targeted mice lacking these receptors exhibit distinct phenotypes, indicating that the two receptors play different roles in health and disease (Deroo and Korach 2006). This has generated a great deal of interest in the therapeutic opportunities afforded by ER-selective ligands. Researchers at Wyeth (Collegeville, PA) have developed selective agonists for ERβ and tested these compounds in several rodent models of human disease (Harris et al. 2003). One such compound, ERB-041, exhibited a beneficial effect in the HLA-B27 transgenic rat model of inflammatory bowel disease as well as the rat adjuvant arthritis model, without affecting ovulation or ovariectomy-induced weight gain, indicating the absence of a classic estrogen-like effect (Harris et al. 2003). These findings pave the way for further studies into the therapeutic potential of selective ER ligands in rheumatoid arthritis and other conditions in which estrogen is thought to play a role.

Acknowledgements. This work was supported by the Arthritis Research Campaign (ARC) of Great Britain.

References

Afuwape AO, Kiriakidis S, Paleolog EM (2002) The role of the angiogenic molecule VEGF in the pathogenesis of rheumatoid arthritis. Histol Histopathol 17:961–72

Albano SA, Santana-Sahagun E, Weisman MH (2001) Cigarette smoking and rheumatoid arthritis. Semin Arthritis Rheum 31:146–59

Alvaro-Garcia JM, Zvaifler NJ, Brown CB, Kaushansky L, Firestein GS (1991) Cytokines in chronic inflammatory arthritis. VI. Analysis of the synovial cells involved in granulocyte-macrophage colony stimulating factor production and gene expression in rheumatoid arthritis and its regulation by IL-1 and TNFα. J Immunol 146:3365–3371

Anonymous (1978) Reduction in incidence of rheumatoid arthritis associated with oral contraceptives. Royal College of General Practitioners' Oral Contraception Study. Lancet 1:569–571

Ballara S, Taylor PC, Reusch P, Marme D, Feldmann M, Maini RN, Paleolog EM (2001) Raised serum vascular endothelial growth factor levels are associated with destructive change in inflammatory arthritis. Arthritis Rheum 44:2055–2064

Banerjee S, Wei BY, Hillman K, Luthra HS, David CS (1988) Immunosuppression of collagen-induced arthritis in mice with an anti-IL-2 receptor antibody. J Immunol 141:1150–1154

Bottazzo GF, Pujol-Borrell R, Hanafusa T, Feldmann M (1983) Role of aberrant HLA-DR expression and antigen presentation in induction of endocrine autoimmunity. Lancet 2:1115–1119

Brennan FM, Chantry D, Jackson A, Maini R, Feldmann M (1989) Inhibitory effect of TNFα antibodies on synovial cell interleukin-1 production in rheumatoid arthritis. Lancet 2:244–247

Brennan FM, Zachariae CO, Chantry D, Larsen CG, Turner M, Maini RN, Matsushima K, Feldmann M (1990) Detection of interleukin 8 biological activity in synovial fluids from patients with rheumatoid arthritis and production of interleukin 8 mRNA by isolated synovial cells. Eur J Immunol 20:2141–2144

Brennan FM, Gibbons DL, Cope AP, Katsikis P, Maini RN, Feldmann M (1995) TNF inhibitors are produced spontaneously by rheumatoid and osteoarthritic synovial joint cell cultures: evidence of feedback control of TNF action. Scand J Immunol 42:158–165

Brennan P, Bankhead C, Silman A, Symmons D (1997) Oral contraceptives and rheumatoid arthritis: results from a primary care-based incident case-control study. Semin Arthritis Rheum 26:817–823

Buchan G, Barrett K, Turner M, Chantry D, Maini RN, Feldmann M (1988) Interleukin-1 and tumour necrosis factor mRNA expression in rheumatoid arthritis: prolonged production of IL-1 alpha. Clin Exp Immunol 73:449–455

Butler DM, Maini RN, Feldmann M, Brennan FM (1995) Modulation of proinflammatory cytokine release in rheumatoid synovial membrane cell cultures. Comparison of monoclonal anti TNF-alpha antibody with the interleukin-1 receptor antagonist. Eur Cytokine Netw 6:225–230

Buzas EI, Hollo K, Rubliczky L, Garzo M, Nyirkos P, Glant TT (1993) Effect of pregnancy on proteoglycan-induced progressive polyarthritis in BALB/c mice: remission of disease activity. Clin Exp Immunol 94:252–260

Choy EH, Kinsley GH, Panayi GS (1998) Monoclonal antibody therapy in rheumatoid arthritis. Br J Rheumatol 37:484–490

Chu CQ, Field M, Abney E, Zheng RQ, Allard S, Feldmann M, Maini RN (1991) Transforming growth factor-beta 1 in rheumatoid synovial membrane and cartilage/pannus junction. Clin Exp Immunol 86:380–386

Cohen S, Katsikis PD, Chu CQ, Thmssen H, Webb LMC, Maini RN, Londei M, Feldmann M (1995) High IL-10 production by the activated T cell population within the rheumatoid synovial membrane. Arthritis Rheum 38:946–952

Cooper SM, Sriram S, Ranges GE (1988) Suppression of murine collagen-induced arthritis with monoclonal anti-Ia antibodies and augmentation with IFN-gamma. J Immunol 141:1958–1962

Cope AP (2003) Exploring the reciprocal relationship between immunity and inflammation in chronic inflammatory arthritis. Rheumatology (Oxford) 42:716–731

Cope AP, Aderka D, Doherty M, Engelmann H, Gibbons D, Jones AC, Brennan FM, Maini RN, Wallach D, Feldmann M (1992) Increased levels of soluble tumor necrosis factor receptors in the sera and synovial fluid of patients with rheumatic diseases. Arthritis Rheum 35:1160–1169

Cope AP, Londei M, Chu NR, Cohen SB, Elliott MJ, Brennan FM, Maini RN, Feldmann M (1994) Chronic exposure to tumor necrosis factor (TNF) in vitro impairs the activation of T cells through the T cell receptor/CD3 complex; reversal in vivo by anti-TNF antibodies in patients with rheumatoid arthritis. J Clin Invest 94:749–760

Da Silva JA, Spector TD (1992) The role of pregnancy in the course and aetiology of rheumatoid arthritis. Clin Rheumatol 11:189–194

Deleuran BW, Chu CQ, Field M, Brennan FM, Katsikis P, Feldmann M, Maini RN (1992) Localization of interleukin-1α, type 1 interleukin-1 receptor and interleukin-1 receptor antagonist in the synovial membrane and cartilage/pannus junction in rheumatoid arthritis. Br J Rheumatol 31:801–809

Deroo BJ, Korach KS (2006) Estrogen receptors and human disease. J Clin Invest 116:561–570

Drossaers-Bakker KW, Zwinderman AH, van Zeben D, Breedveld FC, Hazes JM (2002) Pregnancy and oral contraceptive use do not significantly influence outcome in long term rheumatoid arthritis. Ann Rheum Dis 61:405–408

Ehrenstein MR, Evans JG, Singh A, Moore S, Warnes G, Isenberg DA, Mauri C (2004) Compromised function of regulatory T cells in rheumatoid arthritis and reversal by anti-TNFalpha therapy. J Exp Med 200:277–285

Elliott MJ, Maini RN, Feldmann M, Long-Fox A, Charles P, Katsikis P, Brennan FM, Walker J, Bijl H, Ghrayeb J, Woody JN (1993) Treatment of rheumatoid arthritis with chimeric monoclonal antibodies to tumour necrosis factor α. Arthritis Rheum 36:1681–1690

Elliott MJ, Maini RN, Feldmann M, Kalden JR, Antoni C, Smollen JS, Leeb B, Breedfeld FC, Macfarlane JD, Bijl H, Woody JN (1994a) Treatment with a chimaeric monoclonal antibody to tumour necrosis factor α suppresses disease activity in rheumatoid arthritis: results of a multi-centre, randomised, double blind trial. Lancet 344:1105–1110

Elliott MJ, Maini RN, Feldmann M, Long-Fox A, Charles P, Bijl H, Woody JN (1994b) Repeated therapy with a monoclonal antibody to tumour necrosis factor α in patients with rheumatoid arthritis. Lancet 344:1125–1127

Feldmann M, Maini RN (2001) Anti-TNFα therapy in rheumatoid arthritis: what have we learned? Annu Rev Immunol 19:163–196

Feldmann M, Brennan FM, Maini RN (1996) Role of cytokines in rheumatoid arthritis. Ann Rev Immunol 14:397–440

Feldmann M, Brennan F, Williams RO, Maini RN (2003) Definition of TNFα as a therapeutic target for rheumatoid arthritis. In: Moreland LW, Emery P (eds) TNFα inhibition in the treatment of rheumatoid arthritis. Martin Dunitz, London, pp 1–22

Goldschmidt TJ, Holmdahl R (1991) Anti-T cell receptor antibody treatment of rats with established autologous collagen-induced arthritis: suppression of arthritis without reduction of anti-type II collagen autoantibody levels. Eur J Immunol 21:1327–1330

Hanafusa T, Pujol-Borrell R, Chiovato L, Russell RCG, Doniach D, Bottazzo GF (1983) Aberrant expression of HLA-DR antigen on thyrocytes in Graves' disease: relevance for autoimmunity. Lancet ii:1111–1115

Harris HA, Albert LM, Leathurby Y, Malamas MS, Mewshaw RE, Miller CP, Kharode YP, Marzolf J, Komm BS, Winneker RC, Frail DE, Henderson RA, Zhu Y, Keith JC Jr (2003) Evaluation of an estrogen receptor-beta agonist in animal models of human disease. Endocrinology 144:4241–4249

Harrison BJ, Silman AJ, Wiles NJ, Scott DG, Symmons DP (2001) The association of cigarette smoking with disease outcome in patients with early inflammatory polyarthritis. Arthritis Rheum 44:323–330

Hasler F, van de Putte L, Baudin M, Lodin E, Durrwell L, McAuliffe T, van de Auwera P (1996) Chronic TNF neutralization (up to 1 year) by lenercept (TNFR 55 IgG1, Ro 45–2081) in patients with rheumatoid arthritis: results from open label extension of a double blind single-dose phase I study. Arthritis Rheum 39:S243

Hermann JA, Hall MA, Maini RN, Feldmann M, Brennan FM (1998) Important immunoregulatory role of interleukin-11 in the inflammatory process in rheumatoid arthritis. Arthritis Rheum 41:1388–1397

Hirahara F, Wooley PH, Luthra HS, Coulam CB, Griffiths MM, David CS (1986) Collagen-induced arthritis and pregnancy in mice: the effects of pregnancy on collagen-induced arthritis and the high incidence of infertility in arthritic female mice. Am J Reprod Immunol Microbiol 11:44–54

Hirohata S, Sakakibara J (1999) Angioneogenesis as a possible elusive triggering factor in rheumatoid arthritis. Lancet 353:1331

Hopkins SJ, Meager A (1988) Cytokines in synovial fluid: II. The presence of tumour necrosis factor and interferon. Clin Exp Immunol 73:88–92

Hutchinson D, Shepstone L, Moots R, Lear JT, Lynch MP (2001) Heavy cigarette smoking is strongly associated with rheumatoid arthritis (RA), particularly in patients without a family history of RA. Ann Rheum Dis 60:223–227

Janossy G, Panayi G, Duke O, Bofill M, Poulter LW, Goldstein G (1981) Rheumatoid arthritis: a disease of T-lymphocyte/macrophage immunoregulation. Lancet 2:839–842

Jansson L, Holmdahl R (1989) Oestrogen induced suppression of collagen arthritis. IV: Progesterone alone does not affect the course of arthritis but enhances the oestrogen-mediated therapeutic effect. J Reprod Immunol 15:141–150

Jansson L, Holmdahl R (1992) Oestrogen-induced suppression of collagen arthritis; 17β-oestradiol is therapeutically active in normal and castrated F1 hybrid mice of both sexes. Clin Exp Immunol 89:446–451

Jansson L, Holmdahl R (2001) Enhancement of collagen-induced arthritis in female mice by estrogen receptor blockage. Arthritis Rheum 44:2168–2175

Jansson L, Mattsson A, Mattsson R, Holmdahl R (1990) Estrogen induced suppression of collagen arthritis. V: Physiological level of estrogen in DBA/1 mice is therapeutic on established arthritis, suppresses anti-type II collagen T-cell dependent immunity and stimulates polyclonal B-cell activity. J Autoimmun 3:257–270

Jansson L, Olsson T, Holmdahl R (1994) Estrogen induces a potent suppression of experimental autoimmune encephalomyelitis and collagen-induced arthritis in mice. J Neuroimmunol 53:203–207

Karlson EW, Lee IM, Cook NR, Manson JE, Buring JE, Hennekens CH (1999) A retrospective cohort study of cigarette smoking and risk of rheumatoid arthritis in female health professionals. Arthritis Rheum 42:910–917

Keffer J, Probert L, Cazlaris H, Georgopoulos S, Kaslaris E, Kioussis D, Kollias G (1991) Transgenic mice expressing human tumour necrosis factor: a predictive genetic model of arthritis. EMBO J 10:4025–4031

Klareskog L, Forsum U, Scheynius A, Kabelitz D, Wigzell H (1982) Evidence in support of a self-perpetuating HLA-DR-dependent delayed-type cell reaction in rheumatoid arthritis. Proc Natl Acad Sci USA 79:3632–3636

Londei M, Lamb JR, Bottazzo GF, Feldmann M (1984) Epithelial cells expressing aberrant MHC class II determinants can present antigen to cloned human T cells. Nature 312:639–641

Londei M, Bottazzo GF, Feldmann M (1985) Human T-cell clones from autoimmune thyroid glands: specific recognition of autologous thyroid cells. Science 228:85–89

Malyak M, Swaney RE, Arend WP (1993) Levels of synovial fluid interleukin-1 receptor antagonist in rheumatoid arthritis and other arthropathies. Arthritis Rheum 36:781–789

Mathers D, Russell A (1990) Adjuvant arthritis in the rat during pregnancy. Clin Exp Rheumatol 8:289–292

Mattsson R, Mattsson A, Holmdahl R, Whyte A, Rook GA (1991) Maintained pregnancy levels of oestrogen afford complete protection from post-partum exacerbation of collagen-induced arthritis. Clin Exp Immunol 85:41–47

Miotla J, Maciewicz R, Kendrew J, Feldmann M, Paleolog E (2000) Treatment with soluble VEGF receptor reduces disease severity in murine collagen-induced arthritis. Lab Invest 80:1195–1205

Moreland LW, Baumgartner SW, Schiff MH, Tindall EA, Fleischmann RM, Weaver AL, Ettlinger RE, Cohen S, Koopman WJ, Mohler K, Widmer MB, Blosch CM (1997a) Treatment of rheumatoid arthritis with a recombinant human tumor necrosis factor receptor (p75)-Fc fusion protein. N Engl J Med 337:141–147

Moreland LW, Heck LW, Koopman WJ (1997b) Biologic agents for treating rheumatoid arthritis. Arthritis Rheum 40:397–409

Nelson JL, Ostensen M (1997) Pregnancy and rheumatoid arthritis. Rheum Dis Clin North Am 23:195–212

O'Garra A, Murphy K (1993) T-cell subsets in autoimmunity. Curr Opin Immunol 5:880–886

Paleolog EM, Young S, Stark AC, McCloskey RV, Feldmann M, Maini RN (1998) Modulation of angiogenic vascular endothelial growth factor by tumor necrosis factor alpha and interleukin-1 in rheumatoid arthritis. Arthritis Rheum 41:1258–1265

Piguet PF, Grau GE, Vesin C, Loetscher H, Gentz R, Lesslauer W (1992) Evolution of collagen arthritis in mice is arrested by treatment with anti-tumour necrosis factor (TNF) antibody or a recombinant soluble TNF receptor. Immunology 77:510–514

Powrie F, Correa-Oliveira R, Mauze S, Coffman RL (1994) Regulatory interactions between CD45RBhigh and CD45RBlow CD4+ T cells are important for the balance between protective and pathogenic cell-mediated immunity. J Exp Med 179:589–600

Ranges GE, Sriram S, Cooper SM (1985) Prevention of type II collagen-induced arthritis by in vivo treatment with anti-L3T. J Exp Med 162:1105–1110

Rooney M, Condell D, Quinlan W, Daly L, Whelan A, Feighery C, Bresnihan B (1988) Analysis of the histologic variation of synovitis in rheumatoid arthritis. Arthritis Rheum 31:956–963

Sakaguchi S, Sakaguchi N, Asano M, Itoh M, Toda M (1995) Immunologic self-tolerance maintained by activated T cells expressing IL-2 receptor alpha-chains (CD25) Breakdown of a single mechanism of self-tolerance causes various autoimmune diseases. J Immunol 155:1151–1164

Sheehan KC, Ruddle NH, Schreiber RD (1989) Generation and characterization of hamster monoclonal antibodies that neutralize murine tumor necrosis factors. J Immunol 142:3884–3893

Shevach EM (2000) Regulatory T cells in autoimmmunity. Annu Rev Immunol 18:423–449

Silman AJ, Pearson JE (2002) Epidemiology and genetics of rheumatoid arthritis. Arthritis Res 4(Suppl 3):S265–S272

Silman A, Kay A, Brennan P (1992) Timing of pregnancy in relation to the onset of rheumatoid arthritis. Arthritis Rheum 35:152–155

Silman AJ, MacGregor AJ, Thomson W, Holligan S, Carthy D, Farhan A, Ollier WE (1993) Twin concordance rates for rheumatoid arthritis: results from a nationwide study. Br J Rheumatol 32:903–907

Sivakumar B, Harry LE, Paleolog EM (2004) Modulating angiogenesis: more vs less. JAMA 292:972–977

Sumariwalla PF, Cao Y, Wu HL, Feldmann M, Paleolog EM (2003) The angiogenesis inhibitor protease-activated kringles 1–5 reduces the severity of murine collagen-induced arthritis. Arthritis Res Ther 5:R32–R39

Symons JA, Wood NC, Di Giovine FS, Duff GW (1988) Soluble IL-2 receptor in rheumatoid arthritis. Correlation with disease activity IL-1 and IL-2 inhibition. J Immunol 141:2612–2618

Taylor PC, Peters AM, Paleolog E, Chapman PT, Elliott MJ, McCloskey R, Feldmann M, Maini RN (2000) Reduction of chemokine levels and leukocyte traffic to joints by tumor necrosis factor alpha blockade in patients with rheumatoid arthritis. Arthritis Rheum 43:38–47

Thorbecke GJ, Shah R, Leu CH, Kuruvilla AP, Hardison AM, Palladino MA (1992) Involvement of endogenous tumor necrosis factor α and transforming growth factor β during induction of collagen type II arthritis in mice. Proc Natl Acad Sci USA 89:7375–7379

Todd I, Londei M, Feldmann M, Bottazzo GF (1985) Role of MHC class II molecules and of autoantigens in the pathogenesis of human autoimmune endocrine diseases. In: Milgrom F, Abeyounis CJ, Albini B (eds) Antibodies: protective, destructive and regulatory role. Karger, Basel, pp 66–76

Van Amelsfort JM, Jacobs KM, Bijlsma JW, Lafeber FP, Taams LS (2004) CD4(+)CD25(+) regulatory T cells in rheumatoid arthritis: differences in the presence, phenotype, and function between peripheral blood and synovial fluid. Arthritis Rheum 50:2775–2785

Waites GT, Whyte A (1987) Effect of pregnancy on collagen-induced arthritis in mice. Clin Exp Immunol 67:467–476

Weinblatt ME, Kremer JM, Bankhurst AD, Bulpitt KJ, Fleischmann RM, Fox RI, Jackson CG, Lange M, Burge DJ (1999) A trial of etanercept, a recombinant tumor necrosis factor receptor: Fc fusion protein, in patients with rheumatoid arthritis receiving methotrexate. N Engl J Med 340:253–259

Whyte A, Waites GT (1987) Levels of serum amyloid P-component associated with pregnancy and collagen-induced arthritis in DBA/1 (H-2q) mice. J Reprod Immunoly 12:155–159

Williams RO, Whyte A (1989) A comparison of anti-type II collagen antibody titres and isotype profiles in pregnant and virgin DBA/1 mice with collagen-induced arthritis. J Reprod Immunol 15:229–239

Williams RO, Feldmann M, Maini RN (1992) Anti-tumor necrosis factor ameliorates joint disease in murine collagen-induced arthritis. Proc Natl Acad Sci USA 89:9784–9788

Williams RO, Ghrayeb J, Feldmann M, Maini RN (1995) Successful therapy of collagen-induced arthritis with TNF receptor-IgG fusion protein and combination with anti-CD. Immunology 84:433–439

Wooley PH, Dutcher J, Widmer MB, Gillis S (1993) Influence of a recombinant human soluble tumour necrosis factor receptor Fc fusion protein on type II collagen-induced arthritis in mice. J Immunol 151:6602–6607

Xu WD, Firestein GS, Taetle R, Kaushansky K, Zvaifler NJ (1989) Cytokines in chronic inflammatory arthritis. II. Granulocyte-macrophage colony-stimulating factor in rheumatoid synovial effusions. J Clin Invest 83:876–882

Yoshino S, Cleland LG, Mayrhofer G (1991) Treatment of collagen-induced arthritis in rats with a monoclonal antibody against the alpha/beta T cell antigen receptor. Arthritis Rheum 34:1039–1047

The Role of ERα and ERβ in the Prostate: Insights from Genetic Models and Isoform-Selective Ligands

S.J. McPherson, S.J. Ellem(✉), V. Patchev, K.H. Fritzemeier, G.P. Risbridger

Centre for Urology Research, Monash Institute of Medical Research, Monash University, 27-31 Wright Street Clayton, 3168 Victoria, Australia
email: *Stuart.Ellem@med.monash.edu.au*

Note: Some data and images presented in this manuscript were presented in the article entitled "Essential role for estrogen receptor β in stromal-epithelial regulation of prostatic hyperplasia" by McPherson SJ, Ellem SJ, Simpson ER, Patchev V, Fritzemeier KH and Risbridger GP. This paper was accepted and published online in the journal *Endocrinology* on October 26, 2006.

1	Introduction	132
2	Androgens, Estrogens and Aromatase	133
3	Estrogen and Inflammation of the Prostate	134
4	ArKO Mouse and Hormonal Carcinogenesis	135
5	Estrogens as a Beneficial Influence on Prostate Growth and Development	135
6	Failure to Activate ERβ Results in Induction of Prostatic Epithelial Hyperplasia	139
7	Summary	144
	References	145

Abstract. Androgens are known regulators of the growth and differentiation of the prostate gland and are effective during development and maturity as well as in disease. The role of estrogens is less well characterized, but dual direct and indirect actions on prostate growth and differentiation have been demonstrated, facilitated via both ERα and ERβ. Previous studies using animal models to determine the role of ERβ in the prostate have been problematic due to the centrally mediated responses to estrogen administration via ERα that can lower androgen levels and lead to epithelial regression, thereby masking any direct effects on the prostate mediated by ERβ. Our alternate approach was to use the estrogen-deficient aromatase knockout (ArKO) mouse and the method of tissue recombination to provide new insight into estrogen action on prostate growth and pathology. Firstly, utilizing homo- and heterotypic tissue recombinants, we demonstrate that stromal aromatase deficiency results in the induction of hyperplasia in previously normal prostatic epithelium and that this response is the result of local changes to the paracrine interaction between stroma and epithelium. Secondly, using tissue recombination and an ERβ-specific agonist, we demonstrate that the activation of ERβ results in an anti-proliferative response that is not influenced by alterations to systemic androgen levels or activation of ERα. Finally, using intact ArKO mice this study demonstrates that the administration of an ERβ-specific agonist abrogates existing hyperplastic epithelial pathology specifically in the prostate but an ERβ-specific agonist does not. Therefore, in the absence of stromal aromatase gene expression, epithelial proliferation, leading to prostatic hypertrophy and hyperplasia, may result from a combination of androgenic stimulation of proliferation and failed activation of ERβ by locally synthesized estrogens. These data demonstrate essential and beneficial effects of estrogens that are necessary for normal growth of the prostate and distinguish them from those that adversely alter prostate growth and differentiation. This indicates the potential of antiandrogens and SERMS, as opposed to aromatase inhibitors, for the management of prostate hyperplasia and hypertrophy.

1 Introduction

The prostate gland is located at the neck of the bladder and functions in men to provide secretions that contribute to seminal fluid. The secretions arise from the epithelial secretory cells lining the ducts of this glandular tissue and the ducts themselves are surrounded by stroma. The epithelial–stromal cell interactions, as well as endocrine hormones, are integral to normal development, growth and function of the prostate gland (Cunha et al. 2004).

The human prostate gland undergoes several phases of growth and differentiation, including during fetal and pubertal life as well as in older men. With a doubling time of 2.75 years, the gland grows from 2 g in prepubertal boys to 20 g in mature young men, under the influence of rising androgens. When maturity is reached, the prostate gland becomes growth-quiescent, despite retaining levels of androgens in the circulation comparable to those that initiated prostate maturation. However, after a period of stability, androgen levels decline in older men, yet growth of the prostate gland is resumed and it can more than double in size. Therefore, although androgens are essential and required for prostate growth, they are not the sole regulators or determinants of prostate growth, and regulation of prostate growth and differentiation is far more complex than being controlled by androgens alone. It is in this context that the importance of stromal–epithelial cell interactions will be considered.

Prostate disease in men can be benign or malignant and typically both pathologies will emerge together upon aging. Benign prostatic hyperplasia (BPH) or nodular hyperplasia is due to various combinations of stromal and glandular hyperplasia. On the other hand, prostate cancer (PCa) is predominantly identifiable as adenocarcinoma of the prostate gland. It is generally believed the etiologies of the two diseases are separate and independent, although both BPH and PCa are hormone-dependent diseases.

2 Androgens, Estrogens and Aromatase

The changing ratio of androgens to estrogens is believed to contribute to the emergence of prostate pathologies in older men. Whereas androgens decline upon aging (Huggins and Hodges 1941; Vermeulen et al. 1972; Wu and Gu 1987), the levels of estradiol remain unchanged or increase with age (Vermeulen et al. 1972; Gray et al. 1991a,b; Culig et al. 1993; Baulieu 2002; Vermeulen et al. 2002). Consequently, there is a significant change to the ratio of estradiol to testosterone that is temporally related to the onset of prostate disease, suggesting an adverse role for estrogens in the etiology of PCa and BPH.

Peripheral tissue steroid synthesis may be extremely important in determining the hormonal milieu. Indeed, adipose tissue is one of the major sites wherein androgens are metabolized to estrogens, and as men tend to exhibit a greater degree of adiposity as they age, this is an obvious way through which estrogens may rise with aging (Zumoff et al. 1982; Griffiths 2000; Vermeulen et al. 2002). In addition, there is considerable evidence to show that the prostate itself is a site of estrogen production. It is well known that androgens reaching the prostate gland are reduced by the 5α reductase enzyme to the more potent 5α-dihydrotestosterone (DHT) that activates the prostatic androgen receptor. However, testosterone can also be converted to estrogen by the aromatase enzyme. Located in the stromal tissue of the normal prostate gland, aromatase converts androgens to estrogens which subsequently play a critical role in regulating prostate growth and differentiation in normal and diseased tissue (Ellem et al. 2004). Both estrogen receptor subtypes are identifiable within the stroma and epithelia, confirming that the prostate is a target tissue for estrogens, as well as androgens.

In the normal human prostate, the aromatase enzyme is localized solely to the stroma and gene expression is regulated by the aromatase pII promoter (Ellem et al. 2004). In PCa, the epithelial derived tumor cells now express aromatase, but under the control of additional promoters (1.3 and 1.4) (Ellem et al. 2004). It is of particular note that promoters PII and I.3 may be regulated by inflammatory cytokines, and as estrogens can induce an inflammatory response in the prostate gland, a link between inflammation, estrogens and prostate cancer begins to emerge.

3   Estrogen and Inflammation of the Prostate

There are several sets of data describing estrogen-induced inflammation in this organ. Previously, we have reported that transient up-regulation of estrogen is sufficient to induce an inflammatory response in gonadotrophin and estrogen-deficient (*hpg* and ArKO, respectively) mice (Bianco et al. 2002, 2006). Most recently it was observed that elevated estrogens in aromatase over-expressing (AROM+) mice resulted in chronic prostatic inflammation (Ellem et al., unpublished observations). Hence, estrogens directly cause inflammation in the prostate, and this response is known to be mediated by the ERα receptor subtype (Prins et al. 2001).

These data suggest that induction of inflammation may alter the regulation of aromatase activity, resulting in aberrant local estrogen synthesis that may induce further inflammation, and with unbalanced androgen and estrogen levels, lead to malignancy. Given this implication, we proposed the hypothesis that estrogen deficiency would have a beneficial effect on the development of prostate disease. In order to test this hypothesis, we examined estrogen-deficient ArKO mice.

4 ArKO Mouse and Hormonal Carcinogenesis

It has been previously reported that aromatase knockout (ArKO) mice show prostate hypertrophy and hyperplasia upon aging, which was believed to be due to the lifelong exposure to androgens unopposed by estrogen (McPherson et al. 2001). However, there have been no reports of these animals developing prostate malignancies. Therefore, to test the susceptibility of ArKO mice to the induction of malignancy, a combination of androgens and estrogens were administered to adult ArKO and wild-type (wt) mice. This method of inducing prostatic carcinoma has been successfully used in dogs, mice and most commonly the Noble rat (Ho et al. 1995). Following treatment, the prostate tissues were examined for evidence of malignant or premalignant pathology based on five criteria: altered morphology, up-regulated expression of AR, ERα and PCNA, and loss of E-cadherin expression. Based on these parameters, focal premalignant lesions identifiable as prostatic intraepithelial neoplasia (PIN) could be detected in the prostates of all treated mice. Comparison of ArKO and wt mice showed the incidence of lesions in prostates from estrogen deficiency was approximately 50% of that seen in wt animals (Ricke et al., unpublished observations). This finding adversely implicates estrogen, and aromatase activity, in the development of premalignant lesions of the prostate, ultimately leading to malignancy.

5 Estrogens as a Beneficial Influence on Prostate Growth and Development

In contrast to the adverse responses to estrogen described above, epidemiological studies implicate estrogens or estrogenic substances (e.g. phytoestrogens) as being beneficial in the prevention of prostate disease. Laboratory studies have demonstrated that phytoestrogens suppress tumor cell growth and may have potential therapeutic potential in the prevention of prostate disease (Adlercreutz et al. 1995, 2000; Stephens 1997; Mentor-Marcel et al. 2001). These classes of estrogens preferentially bind the ERβ, rather than the ERα, subtype and are purported to be anti-proliferative, although this biological action remains unproven (Kuiper et al. 1998). Within the male reproductive tract, the prostate epithelial cells expresses very high levels of ERβ yet, to date, it has been difficult to demonstrate a direct anti-proliferative response to estrogen in this (or any other) tissue, because administration of exogenous estrogen to a male will cause a reduction in androgens induce apoptosis, reduce proliferation and cause prostatic epithelial atrophy. Thus the putative anti-proliferative effects of ERβ activation by exogenous estrogen cannot be distinguished from the effects of reducing systemic androgens.

In this context, we re-evaluated the role of cell–cell signaling in the prostate gland, since the prostatic micro-environment is a critical regulator of growth and estrogens are key contributors to the prostatic microenvironment.

Based on the expression of aromatase in prostatic stroma (Ellem et al. 2004), we used tissue recombination to determine if estrogen deficiency within the stroma would alter prostatic epithelial cell differentiation and/or proliferation. A similar approach was used to elucidate the role of prostatic stromal AR signaling using *Tfm* mouse tissues (Cunha and Donjacour 1987; Thompson et al. 2000) and more recently, estrogen action via ERα using ERαKO tissues (Risbridger et al. 2001a). The power of this technique to elucidate stromal–epithelial interactions involving estrogen arises from two factors. Firstly, tissue recombinants are grown in intact hosts in which systemic androgen and estrogen levels are normal, and secondly, recombinants composed of different stroma–epithelial tissue types are grown in the same host animals and are therefore exposed to an identical systemic hormonal milieu. Consequently, any differences between the grafts must be due to the perturbation of stromal–epithelial signaling in the grafted tissues and independent of systemic hormones.

Using this method, we examined whether newborn stroma from aromatase-deficient ArKO mice (ArKO-S) was capable of directing and inducing ductal tip epithelia from wt adult mice (wt-E) to become hyperplastic, a phenotype similar to that observed in the ArKO mouse upon maturation of the prostatic epithelium. In addition to a visual examination of the histology of the grafts, we directly compared the degree of epithelial infolding, or branching, in the tissue recombinants as a measure of hyperplasia.

The importance of local stromal–epithelial cell interaction is unequivocal and clearly demonstrated by the comparison of the homotypic tissue recombinants (wt-S+wt-E and ArKO-S+ArKO-E) (Fig. 1a,b). The wt recombinants formed normal prostate tissue as expected, while those derived from ArKO tissues generated prostate tissue with extensive epithelial infolding, characteristic of epithelial hyperplasia previously reported in the prostates of adult ArKO mice (McPherson et al. 2001). Analysis showed that epithelial branching and infolding in the homotypic recombinants was significantly increased, being approximately three times greater in ArKO-S+ArKO-E recombinants compared to wt-S+wt-E recombinants (Fig. 1e). As both types of homotypic recombinant were grafted into the same host animal and exposed to an identical systemic hormonal environment, the difference in epithelial hyperplasia must be attributable to the effects of local estrogen deficiency and perturbation of stromal–epithelial cell interactions.

The pivotal role of the stroma in the induction of epithelial hyperplasia was evident from the comparison of the heterotypic tissue recombinants (ArKO-S+wt-E and wt-S+ArKO-E; Fig. 1c and d, respectively) with the homotypic recombinants. Epithelial hyperplasia in the ArKO-S+wt-E recombinant was significantly different to that of the wt-S+wt-E control recombinant (Fig. 1a) and was comparable to that of the ArKO-S+ArKO-E grafts (Fig. 1b). This demonstrates that the stroma derived from the ArKO mouse is able to initiate epithelial hyperplasia in the normal (wt) epithelia. Interestingly, wt-S+ArKO-E recombinants also display epithelial hyperplasia significantly higher than wt-S+wt-E, and similar to ArKO-S+ArKO-E recombinants (Fig. 1e). The AP from an adult ArKO exhibits significant hyperplasia, and therefore the AP tip used to prepare the wt-S+ArKO-E graft consisted of a piece of tissue that was hyperplastic when it was recombined with the wt stroma. Histological examination of these grafts showed that the hyperplasia persisted and was not altered or reversed when grown in a normal hormonal milieu. The retention of the hyperplastic pathology may be attributed

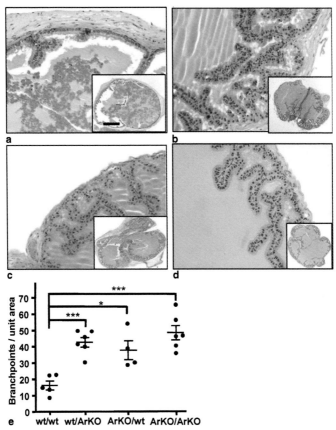

Fig. 1a–e. Formation of homotypic and heterotypic prostate tissue recombinants using ArKO and wt mice. Local stromal–epithelial cell interactions direct prostatic hyperplasia independent of systemic hormones. Tissue recombinants derived from wt-S/wt-E (**a**), ArKO-S/ArKO-E (**b**), ArKO-S/wt-E (**c**) or wt-S/ArKO-E (**d**) from prostate with differing degrees of epithelial hyperplasia as demonstrated by epithelial infolding. **e** The extent of epithelial infolding (estimated as the number of branches per square millimeter), was significantly increased in all homoheterotypic ArKO tissue recombinants compared to wt-S/wt-E controls despite exposure to the same hormonal milieu. Values represent mean ± SEM, $n \geq 4$; ($p < 0.05$); significant difference indicated by different superscripts. *Scale bars* = 100 μm (**a–d**); 400 μm (*insets*)

Table 1 Incidence of AR immunolocalization

Recombinant	Incidence of AR immunolocalization in		
	Epithelium	Stroma	Total
wt-s/wt-e	15.53% ± 0.10	31.57% ± 2.30	18.53% ± 0.24
wt-s/ArKO-e	10.32% ± 4.03	28.62% ± 6.08	13.50% ± 2.68
ArKO-s/wt-e	13.17% ± 3.81	27.02% ± 12.34	14.61% ± 4.37
ArKO-s/ArKO-e	11.47% ± 1.85	24.30% ± 3.28	14.04% ± 1.72

to a failure of the correct signals from wt stroma to reverse adult epithelial hyperplasia or the need to increase the duration of the epithelial exposure to signals from the wt stroma, requiring further time to be effective. Nevertheless, the effect of estrogen deficiency in neonatal stroma alone is sufficient to induce epithelial hyperplasia in tissue recombinants grafted into host mice where normal androgen and estrogen levels are maintained. This provides a new insight to the pivotal role of aberrant stromal–epithelial cell signaling in the onset of epithelial cell hyperplasia.

In order to elucidate the underlying mechanism of aberrant stromal–epithelial signaling, we examined the expression of AR in the recombinant tissues. As demonstrated above, AR is elevated in stroma from neonatal estrogen-deficient ArKO mice. Further examination of the recombinant tissues failed to identify any difference in AR expression (Table 1) or any correlation between AR expression and the induction of hyperplasia in any group of tissue recombinants using either neonatal ArKO or wt stroma (data not shown).

6 Failure to Activate ERβ Results in Induction of Prostatic Epithelial Hyperplasia

Since estrogen signaling via ERβ is postulated to be anti-proliferative, we next investigated the activation of this receptor subtype in the tissue recombinants. As the recombinants are exposed to the same systemic hormones in the host mice, it was possible to study estrogen action without the complication of any differences in serum androgen levels.

The preceding result using homotypic recombinants demonstrate that correct stromal–epithelial signaling of estrogens is a key determinant of prostatic hyperplasia in the ArKO mouse. We did not consider estrogen action via ERα because it has been previously shown to result in epithelial cell proliferation leading to squamous metaplasia (Risbridger et al. 2001b). Squamous metaplasia is a proliferative response that differs from the epithelial hyperplasia observed in the ArKO mice (Risbridger et al. 2001b). Instead, we re-considered the role of ERβ as a possible anti-proliferative influence on prostate growth, despite the lack of biological endpoints for it in this role (Weihua et al. 2001; Imamov et al. 2004). However, there are conflicting data on the developmental expression of ERβ; one group has reported the absence of significant ERβ expression in mice until after the 1st week of life (Omoto et al. 2005), whereas ERβ protein was identified in human and rat fetal tissues (Shapiro et al. 2005; Chrisman and Thomson 2006).

In order to determine if a failure to activate ERβ is due to the absence of conversion of androgens to estrogens in the aromatase-deficient stroma, we again used the tissue recombination model. Based on the above results showing induction of epithelial hyperplasia when neonatal stroma from ArKO mice (ArKO-S) is recombined with normal epithelial cells of ductal tips (wt-E), we subsequently treated adult host mice bearing homotypic ArKO-S/ArKO-E and wt-S/wt-E with an Erβ-specific agonist. Compared to ArKO-S/ArKO-E recombinants treated with vehicle (Fig. 2b), treatment with ERβ agonist abrogated epithelial hyperplasia in ArKO-S/ArKO-E recombinants (Fig. 2c), so that tissues were indistinguishable from wt-S/wt-E tissue recombinants (Fig. 2a). Prostatic epithelial cell proliferation was significantly reduced within each of these recombinants, as determined by localization of PCNA expression (Fig. 2d). There was no evidence of an increase in apoptosis, which was almost undetectable (data not shown).

These data unequivocally demonstrate the essential role of locally metabolized estrogen in the stroma regulating epithelial cell differentiation and proliferation during development. Additionally, the anti-proliferative response mediated via ERβ occurs in contrast to the proliferative response to estrogen mediated by ERα, which causes squamous metaplasia (Risbridger et al. 2001b). Hence these data provide new insight into the role of ERβ in prostate development and differentiation

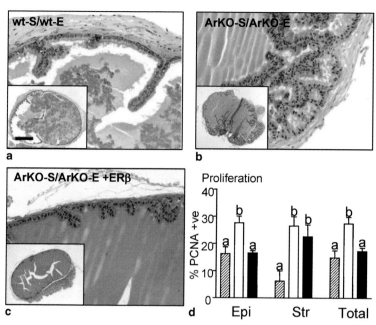

Fig. 2a–d. Estrogen receptor β agonist reduces epithelial hyperplasia in ArKO prostate tissue recombinants. **a** wt-S/wt-E recombinant from vehicle-treated host showing normal epithelium. **b** ArKO-S/ArKO-E recombinant from vehicle-treated host showing epithelial hyperplasia. **c** ArKO-S/ArKO-E recombinant from an ERβ agonist-treated host shows loss of epithelial hyperplasia and is comparable to wt-S/wt-E control. *Scale bars*: 100 μm (**a–d**); 400 μm (*insets*). **d** Proliferation as determined by PCNA expression in ArKO-S/ArKO-E recombinants (*open bars*) was significantly elevated compared to wt-S/wt-E recombinants (*hatched bars*); treatment of ArKO-S/ArKO-E recombinants with ERβ agonist (*solid bars*) significantly reduced the percentage of cells proliferating in epithelium and in total, although stromal proliferation was not altered. Values indicate mean ± SEM, $n \geq 5$, superscripts indicate groups that are not significantly different ($p < 0.05$)

and conclusively prove the anti-proliferative action without any confounding difference in systemic androgen levels. These data show a critical role of stromal aromatase enzyme activity required for the synthesis of local estrogen which prevents epithelial hyperplasia via ERβ signaling.

Prostatic hypertrophy and glandular and/or stromal tissue hyperplasia occur concomitantly in men with BPH and are believed to arise in an environment of unbalanced androgen and estrogen in late life. This study uses a murine model of prostate hypertrophy and hyperplasia to provide new insight into the interactions of these hormones during development. Firstly, it demonstrates that the correct balance between androgens and estrogens is critical for normal development of the prostate gland at the start of life. Secondly, this study identifies an important beneficial role for estrogen during neonatal development; whereas androgens drive cell proliferation and prostate growth in a coordinated manner, locally synthesized estrogen activates ERβ and prevents epithelial hyperplasia via stromal–epithelial interactions.

The previous findings suggest the potential of ERβ agonists for therapeutic treatment of benign prostatic hyperplasia. Therefore we have examined whether administration of ERβ-specific agonists could abrogate existing prostatic hyperplasia in adult ArKO mice. Following 6 weeks of systemic administration of a specific ERβ agonist to intact ArKO male mice, serum T levels were not significantly different from control animals (Fig. 3a), nor was seminal vesicle (SV) weight altered (Fig. 3b). however, ventral prostate weight was significantly reduced (Fig. 3c) and histological examination showed a reduction in epithelial hyperplasia,

Fig. 3a–g. Estrogen receptor β agonist reduces prostate prostatic epithelial hyperplasia in intact ArKO mice. **a** Serum T levels were not altered in ArKO mice treated with vehicle control (*P, solid bar*), (ERα 0.3 or 3.0 μg/kg/day; *open bars*) or ERβ agonist (30 or 100 μg/kg/day; *shaded bars*). **b** SV weights were significantly reduced by ERα but not ERβ agonist. **c** VP weights were significantly reduced by ERα (3.0 μg) and ERβ agonist (30 and 100 μg). Compared to normal wt prostates (**d**), ArKO prostates (**e**) demonstrate a hyperplastic epithelial morphology throughout the tissue. **f** ArKO tissue treated with ERβ agonist (100 μg) show epithelial morphology comparable to wt prostate, while ArKO tissue treated with an ERα agonist (**g**; 3.0 μg) shows no reduction in epithelial hyperplasia. Values represent mean ± SEM; * significance $p < 0.05, n \geq 6$ compared to ArKO control. *Scale bar* = 100 μm (*DG*)

ERα and ERβ in the Prostate

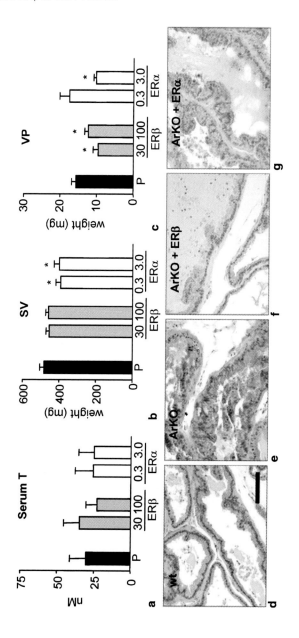

resulting in areas of ArKO tissue (Fig. 3f) that were morphologically indistinguishable from wt controls (Fig. 3d). In contrast, the administration of an ERα agonist reduced SV weight and had variable, dose-dependent effects on prostate weight (Fig. 3c) but failed to reverse the hyperplastic phenotype (Fig. 3g). Of more concern and, consistent with previous demonstrations that ERα mediates adverse inflammatory responses in the prostate gland (Prins et al. 2001), some regions of tissue showed evidence of inflammation and infiltration of inflammatory cells (data not shown). The reduction in SV weight following ERα agonist treatment suggests lower serum androgen levels in ERβ-agonist-treated mice might induce prostatic epithelial atrophy and, together with the observed inflammation, this treatment could lead to prostatic inflammatory atrophy, a pathology linked to premalignancy (De Marzo et al. 1999; Nelson et al. 2004). The data presented confirm the adverse prostatic response to activation of ERα, but demonstrate the beneficial actions of estrogen action via ERβ in reducing prostatic epithelial hyperplasia.

7 Summary

These studies show that estrogens as well as androgens are powerful regulators of prostatic growth and differentiation. The actions of the two classes of steroids are similar but different. Activation of androgen receptor stimulates proliferation and differentiation of the prostatic epithelium; activation of ERα also induces a proliferative response, but of an aberrant nature, leading to inflammation and increased predisposition to malignancy. The failure to activate androgen receptor following castration reduces prostatic growth, whereas the failure to activate ERβ leads to hypertrophy and hyperplasia. Thus these studies underscore the importance of considering the combined effects of androgens and estrogens in the endocrine hormone regulation of the prostate gland. Based on these observations, it would be predicted that the therapeutic utility of hormone-based treatment of prostatic hypertrophy and hyperplasia would require the administration of androgen and ERα antagonists plus ERβ agonist.

References

Adlercreutz CH, Goldin BR, Gorbach SL et al (1995) Soybean phytoestrogen intake and cancer risk. J Nutr 125(3 Suppl):757S–770S

Adlercreutz H, Mazur W, Bartels P et al (2000) Phytoestrogens and prostate disease. J Nutr 130:658S–659S

Baulieu EE (2002) Androgens and aging men. Mol Cell Endocrinol 198:41–49

Bianco JJ, Handelsman DJ, Pedersen JS et al (2002) Direct response of the murine prostate gland and seminal vesicles to estradiol. Endocrinology 143:4922–4933

Bianco JJ, McPherson SJ, Wang H et al (2006) Transient neonatal estrogen exposure to estrogen deficient mice (Aromatase knockout) reduces prostate weight and induces inflammation in late life. Am J Pathol 168:1869–1878

Chrisman H, Thomson AA (2006) Regulation of urogenital smooth muscle patterning by testosterone and estrogen during prostatic induction. Prostate 66:696–707

Culig Z, Hobisch A, Cronauer MV et al (1993) Mutant androgen receptor detected in an advanced-stage prostatic carcinoma is activated by adrenal androgens and progesterone. Mol Endocrinol 7:1541–1550

Cunha GR, Donjacour A (1987) Stromal-epithelial interactions in normal and abnormal prostatic development. Prog Clin Biol Res 239:251–272

Cunha GR, Cooke PS, Kurita T (2004) Role of stromal-epithelial interactions in hormonal responses. Arch Histol Cytol 67:417–434

De Marzo AM, Marchi VL, Epstein JL et al (1999) Proliferative inflammatory atrophy of the prostate: implications for prostatic carcinogenesis. Am J Pathol 155:1985–1992

Ellem SJ, Schmitt JF, Pedersen JS et al (2004) Local aromatase expression in human prostate is altered in malignancy. J Clin Endocrinol Metab 89:2434–2441

Gray A, Feldman HA, McKinlay JB et al (1991a) Age, disease, and changing sex hormone levels in middle-aged men: results of the Massachusetts Male Aging Study. J Clin Endocrinol Metab 73:1016–1025

Gray A, Berlin JA, McKinlay JB et al (1991b) An examination of research design effects on the association of testosterone and male aging: results of a meta-analysis. J Clin Epidemiol 44:671–684

Griffiths K (2000) Estrogens and prostatic disease. International Prostate Health Council Study Group. Prostate 45:87–100

Ho S, Leav I, Merk FB et al (1995) Induction of atypical hyperplasia, apoptosis, and type II estrogen-binding sites in the ventral prostates of Noble rats treated with testosterone and pharmacologic doses of estradiol-17 beta. Lab Invest 73:356–365

Huggins C, Hodges CV (1941) Studies on prostatic cancer. The effect of castration, of estrogen and of androgen interaction on serum phosphatases in metastatic carcinoma of the prostate. Cancer Res 1:293–297

Imamov O, Morani A, Shim GJ et al (2004) Estrogen receptor beta regulates epithelial cellular differentiation in the mouse ventral prostate. Proc Natl Acad Sci USA 101:9375–9380

Kuiper GG, Lemmen JG, Carlsson B et al (1998) Interaction of estrogenic chemicals and phytoestrogens with estrogen receptor beta. Endocrinology 139:4252–4263

McPherson S, Wang H, Jones ME et al (2001) Elevated androgens and prolactin in aromatase deficient (ArKO) mice cause enlargement but not malignancy of the prostate gland. Endocrinology 142:2458–2467

Mentor-Marcel R, Lamartiniere CA, Eltoum IE et al (2001) Genistein in the diet reduces the incidence of poorly differentiated prostatic adenocarcinoma in transgenic mice (TRAMP). Cancer Res 61:6777–6782

Nelson WG, De Marzo AM, De Weese TL et al (2004) The role of inflammation in the pathogenesis of prostate cancer. J Urol 172:S6–S11; discussion S11–S2

Omoto Y, Imamov O, Warner M et al (2005) Estrogen receptor alpha and imprinting of the neonatal mouse ventral prostate by estrogen. Proc Natl Acad Sci USA 102:1484–1489

Prins GS, Birch L, Couse JF et al (2001) Estrogen imprinting of the developing prostate gland is mediated through stromal estrogen receptor alpha: studies with alphaERKO and betaERKO mice. Cancer Res 61:6089–6097

Risbridger G, Wang H, Young P et al (2001a) Evidence that epithelial and mesenchymal estrogen receptor-alpha mediates effects of estrogen on prostatic epithelium. Dev Biol 229:432–442

Risbridger GP, Wang H, Frydenberg M et al (2001b) The metaplastic effects of estrogen on mouse prostate epithelium: proliferation of cells with basal cell phenotype. Endocrinology 142:2443–2450

Shapiro E, Huang H, Masch RJ et al (2005) Immunolocalization of estrogen receptor alpha and beta in human fetal prostate. J Urol 174:2051–2053

Stephens FO (1997) Phytoestrogens and prostate cancer: possible preventive role. Med J Aust 167:138–140

Thompson TC, Timme TL, Park SH et al (2000) Mouse prostate reconstitution model system: a series of in vivo and in vitro models for benign and malignant prostatic disease. Prostate 43:248–254

Vermeulen A, Rubens R, Verdonck L (1972) Testosterone secretion and metabolism in male senescence. J Clin Endocrinol Metab 34:730–735

Vermeulen A, Kaufman JM, Goemaere S et al (2002) Estradiol in elderly men. Aging Male 5:98–102

Weihua Z, Makela S, Andersson LC et al (2001) A role for estrogen receptor beta in the regulation of growth of the ventral prostate. Proc Natl Acad Sci USA 98:6330–6335

Wu JP, Gu FL (1987) The prostate 41–65 years post castration. An analysis of 26 eunuchs. Chin Med J (Engl) 100:271–272

Zumoff B, Levin J, Strain GW et al (1982) Abnormal levels of plasma hormones in men with prostate cancer: evidence toward a two-disease theory. Prostate 3:579–588

Ernst Schering Foundation Symposium Proceedings, Vol. 1, pp. 149–161
DOI 10.1007/2789_2006_021
© Springer-Verlag Berlin Heidelberg
Published Online: 4 May 2007

Preclinical Characterization of Selective Estrogen Receptor β Agonists: New Insights into Their Therapeutic Potential

H.A. Harris(✉)

Wyeth Research, Women's Health Research Institute, 500 Arcola Rd, RN-3163, 19426 Collegeville PA, USA
email: harrish@wyeth.com

1	Introduction and Review of Compounds Discussed	150
2	ERβ Agonists Lack Classic Estrogenic Effects	151
3	Improving Intestinal Function: A Common Theme Among Three Vivo Efficacy Models	153
3.1	HLA-B27 Transgenic Rat	153
3.2	Mdr1aKO Mouse Model of Colitis	153
3.3	Rodent Models of Sepsis	154
4	Endometriosis and Inflammatory Pain	155
5	Hypothalamic–Pituitary–Adrenal Axis: A New Area of Investigation	156
6	Summary and Future Directions	157
References		158

Abstract. It has now been over 10 years since Jan-Ake Gustafsson revealed the existence of a second form of the estrogen receptor (ERβ) at a 1996 Keystone Symposium. Since then, substantial success has been made in distinguishing its potential biological functions from the previously known form (now called ERα) and how it might be exploited as a drug target. Subtype selective agonists have been particularly useful in this regard and suggest that ERβ agonists may be useful for a variety of clinical applications without triggering classic estrogenic side effects such as uterine stimulation. These applications include inflammatory bowel disease, rheumatoid arthritis, endometriosis, and sepsis. This manuscript will summarize illustrative data for three ERβ selective agonists, ERB-041, WAY-202196, and WAY-200070.

1 Introduction and Review of Compounds Discussed

Estrogens are classically described as ligand-activated transcription factors that modulate gene transcription via two intracellular receptors, ERα and ERβ. Although other mechanisms of signaling have been described, this path is the best described. ERα was the first ER cloned and a knockout mouse was made approximately 13 years ago. Although not all aspects of estrogen biology were neatly solved by these discoveries, the field generally accepted that there was but a single ER. When ERβ was unexpectedly discovered in a rat prostate cDNA library 10 years ago, it rejuvenated the field of estrogen research with the new goal of attributing estrogen's panoply of effects to the appropriate ER. A number of approaches were taken to investigate this question, including examining receptor distribution, in vitro activity/interaction studies, construction of knockout mice, and design of selective agonists/antagonists. Given the similarity of ERα and ERβ ligand-binding pockets, it was a particular challenge to develop highly selective ligands. In fact, although a variety of selective agonists have been developed (Veeneman 2005), highly selective antagonists have not been designed.

Data from three selective agonists synthesized by Wyeth Research (Malamas et al. 2004; Mewshaw et al. 2005) will be presented in an effort to outline our current understanding of ERβ biology. The majority of material discussed here was presented at the Ernst Schering Symposium on Tissue-Specific Estrogen Action in March 2006. The biology of other ER selective agonists (both ERα and ERβ) has been recently reviewed elsewhere (Harris 2007).

The structure of compounds discussed in this article and their binding affinities (as measured by IC_{50}) are shown in Fig. 1. All three compounds are nonsteroidal agonists and bind to human ERβ ligand binding domain with roughly the same affinity as 17β-estradiol. Their selectivity in this assay varies from approximately 70- to more than 200-fold. In

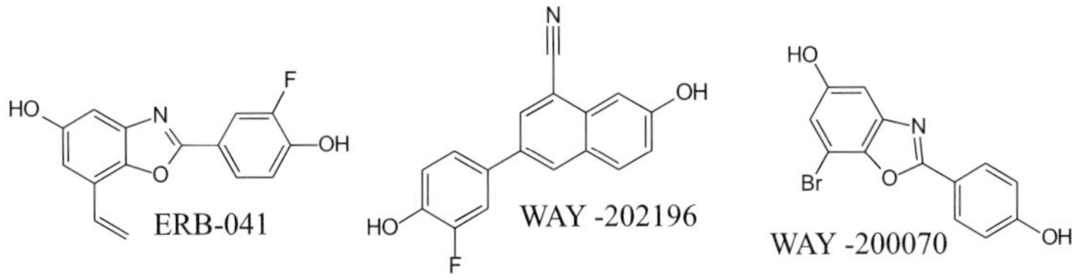

	ERβ IC$_{50}$ (nM) Mean ± SD	ERα IC$_{50}$ (nM) Mean ± SD	Fold selectivity for ERβ
17β-estradiol	3.6 ± 1.6	3.2 ± 1.0	1
ERB-041	5.4 ± 3.7	1216 ± 688	220
WAY-202196	2.7 ± 1.9	210 ± 122	77
WAY-200070	2.0 ± 1.0	155 ± 47	68

Fig. 1. The structure of ERβ agonists discussed in this article, their binding affinity (as measured by IC$_{50}$) and selectivity as assessed using the human ligand-binding domain in a competitive radioligand-binding assay

our experience, these and other structurally diverse ERβ agonists have similar in vivo profiles, although we cannot be certain that these compounds are capable of eliciting all ERβ-mediated effects. As more in vivo data is published on other ERβ selective agonists, the full spectrum of ERβ biology will be elaborated. It should be noted that the activity of other ERβ selective agonists is discussed elsewhere in this volume (see Chaps. 4, 6, 7 and 8).

2 ERβ Agonists Lack Classic Estrogenic Effects

Key to ERβ's attractiveness as a drug target was the expectation that selective agonists would have reduced impact on the uterus and mammary gland. This assumption was supported by tissue distribution studies and the phenotype of the ERα knockout mouse. Indeed, in our hands, these three compounds (as well as many others) are nonuterotrophic in the sexually immature rat. However, given subcutaneously at high doses (~90 mg/kg), to ovariectomized adult rats, WAY-200070 does have some mild uterine stimulatory activity (unpublished observations). As part of our safety assessment of ERB-041 and WAY-202196 in preparation for clinical trials, higher doses have been tested orally in two species and no uterine stimulatory activity has been seen.

The mouse mammary gland responds to the combination of an estrogen and a progestin by elaboration/development of ducts and formation of lobuloalveolar endbuds. Typically these steroids are administered for approximately 3 weeks in order to see full development of mammary gland morphology. However, to facilitate compound evaluation, a 7-day model was developed (Crabtree et al. 2006). Under this abbreviated regimen, morphological changes still occur, although they are less pronounced than those seen with longer exposure. In addition, we measured defensin β1 mRNA expression. This gene is uniquely upregulated by the combination of estrogen and progestin; neither compound

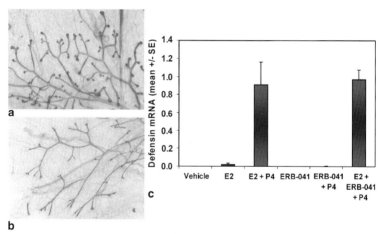

Fig. 2a–c. Activity of ERB-041 in the 7-day mouse mammotrophic assay. a Whole mount images of mammary glands from animals treated with 17β-estradiol (1 μg/kg) + progesterone (30 mg/kg). b Whole mount images of mammary glands from animals treated with ERB-041 (50 mg/kg) + progesterone (30 mg/kg). c Defensin β1 mRNA expression in mammary glands from animals treated with 17β-estradiol (E2, 1 μg/kg), 17β-estradiol (1 μg/kg) + progesterone (30 mg/kg) (E2 + P4), ERB-041 (50 mg/kg), ERB-041 + progesterone (30 mg/kg) (ERB-041 + P4) or 17β-estradiol (1 μg/kg) + progesterone (30 mg/kg) + ERB-041 (50 mg/kg) (E2 + ERB-041 + P4)

alone elevates its expression. ERB-041 and WAY-202196 (50 mg/kg, PO) were evaluated alone and in combination with estradiol, estradiol + progesterone, and progesterone in this model and did not appreciably alter morphology or significantly change defensinβ1 mRNA expression (Fig. 2; Crabtree et al. 2006). Therefore, they are inactive as estrogens, antiestrogens, progestins, or antiprogestins in this model.

ERB-041 and WAY-200070 (10 mg/kg SC) have been evaluated for their ability to prevent bone mineral density loss after ovariectomy in the rat (Harris et al. 2003; Malamas et al. 2004). Unlike estrogens, selective estrogen receptor modulators (Miller 2002) or ERα selective agonists (Harris et al. 2002; Hillisch et al. 2004), these compounds were inactive. Finally, ERB-041 failed to inhibit ovulation in rats (Harris et al. 2003), another point of divergence between ERβ selective agonists and nonsubtype selective estrogens.

3 Improving Intestinal Function: A Common Theme Among Three Vivo Efficacy Models

3.1 HLA-B27 Transgenic Rat

The first in vivo activity seen with these ERβ selective agonists was in a model of inflammatory bowel disease, the HLA-B27 transgenic rat (Harris et al. 2003). These rats experience chronic diarrhea from about 8–10 weeks of age until their death. Daily oral doses of ERB-041 (Harris et al. 2003) and WAY-202196 (Mewshaw et al. 2005) given to male rats rapidly normalized stool character and improved intestinal histology. In this model, doses of 1 mg/kg (PO) and greater were fully effective. The mechanism of action of ERβ is unclear, as is the target cell; ERβ is expressed in the colonic epithelium as well as cells of the immune system, making multiple sites of action possible.

3.2 Mdr1aKO Mouse Model of Colitis

Mdr1a (P-glycoprotein) knockout mice spontaneously develop colitis due to defects in intestinal barrier epithelial function, and males develop earlier and more severe disease than females (Resta-Lenert et al. 2005). Because of the effects seen in the HLA-B27 transgenic rat, a group of male and female mdr1aKO mice were treated orally with either vehicle or WAY-200070 (2 mg/kg) for 10 days, at which time colonic epithelium was removed for functional studies and expression of inflammatory mediators (Barrett and Resta-Lenert, unpublished observations). When placed in Ussing chambers to assess epithelial integrity, colonic epithelium from vehicle-treated mdr1aKO mice had elevated baseline current, indicating a defect in barrier function, and this deficit was significantly reversed when mice were treated with WAY-200070. In addition, when stimulated by forskolin, colonic epithelium from vehicle-treated mice responded poorly to this secretagogue, whereas epithelium from mice treated with WAY-200070 responded normally. Finally, epithelium from vehicle-treated mice had elevated COX-2 and iNOS protein expression and this increase was largely blunted by WAY-200070. Because WAY-200070 has poor systemic exposure upon oral dosing, these data suggest the compound affects the colonic epithelium directly.

3.3 Rodent Models of Sepsis

Sepsis can be generally described as a maladaptive response to infection, and two hallmarks of this disease are epithelial and endothelial barrier dysfunction (Buras et al. 2005; Vincent and Abraham 2005). The advantage that females have over males in conditions of trauma and shock have been well described (Angele et al. 2000). Recent work has specifically focused on the intestine as a target organ for damage and shown that female rats have improved barrier function and mount less of a pro-inflammatory response (as measured by IL-6 and TNFα) than do males after hypoxic or acidic insult (Homma et al. 2005).

Several animal models of sepsis exist, and WAY-202196 has been evaluated in two: *Pseudomonas* infection of the neutropenic rat and mouse cecal ligation and puncture (Cristofaro et al. 2006). In the first model, rats are rendered neutropenic by cyclophosphamide and normal gut flora is disrupted by an antibiotic. They then receive an oral bolus of *Pseudomonas aeruginosa*. WAY-202196 was studied twice. In the first study, rats were dosed on days 4 and 6 after *Pseudomonas* inoculation. In the second study, rats were dosed from days 4–11 after *Pseudomonas* inoculation. In both studies, histological signs of injury were significantly reduced in the intestinal epithelium. Moreover, in the longer-term study, survival was significantly increased by administration of the ERβ selective agonist.

In the second model, puncturing the cecum and expressing a small amount of bowel contents into the peritoneal cavity induces a peritoneal infection. WAY-202196 was administered at the time of surgery, 24 and 48 h afterwards. Consistent with the observation from the rat model, WAY-202196 increased survival, and histological signs of injury were significantly reduced in the ileum. Follow-up studies indicate that the compound is equally effective in males and females, that the minimum effective oral dose of WAY-202196 is 1 mg/kg and that intravenous dosing is as effective as oral dosing (Cristofaro et al. 2006; Opal et al., unpublished data).

4 Endometriosis and Inflammatory Pain

Although endometriosis is undoubtedly an estrogen-responsive disease, it is now appreciated that immune system dysfunction may explain why only a subset of women with retrograde menstruation develop the disease. Because of the anti-inflammatory activity seen with ERB-041 in other models, we evaluated it in a rodent model of endometriosis. The model chosen was a xenograft model using normal human endometrial fragments implanted into nude mice. These tissue fragments adhere, implant, establish a blood supply, and form lesions that are histologically similar to human disease (Bruner et al. 1997; Grummer et al. 2001). Dosing with ERB-041 began approximately 2 weeks after tissue implantation, and continued for about 2 weeks. Spontaneous lesion regression was not seen in vehicle-treated mice, but 40%–75% of mice treated with ERB-041 were completely lesion-free (depending on the study) (Harris et al. 2005). Interestingly, ERB-041 seemed more effective at causing lesion regression when implants were established inside the peritoneal cavity than subcutaneously. As with the other models, the mechanism of action is uncertain; however, because lesions recovered at the end of the study express ERα and not ERβ, the compound is likely acting on the host (e.g., an immunomodulatory effect) rather than on the implants (a pro-apoptotic effect). The fact that ERB-041 is active in a model of this disease illustrates that an ERβ agonist's profile is not just a subset of estradiol's activity or that of other nonselective ER agonists.

Pain is a central feature of several diseases for which ERβ selective agonists have been effective in preclinical models. We examined whether ERB-041 had antinociceptive activity in several models, including a model of inflammatory pain (Leventhal et al. 2006). When prostaglandin E2 is injected into a rat's tail, the tail becomes hypersensitive to warm water. Acute oral administration of ERB-041 (10 mg/kg) can partially reverse this effect, and its action is blocked by the ER antagonist ICI-182780. Similar results are seen when capsaicin is used as the sensitizer. However, ERB-041 was not effective in other models of pain, including postsurgical pain and neuropathic pain. Again, not understanding the mechanism of action impedes an explanation for these patterns of activity.

5 Hypothalamic–Pituitary–Adrenal Axis: A New Area of Investigation

One of the challenges we face regarding our compounds is the ambiguity of their mechanism of action. To date, all the in vivo activities described for this set of ERβ selective agonists relate to inflammation and/or the immune system. These activities may be anti-inflammatory, immunomodulatory, or potentially even immunostimulatory, but thus far, they share this common thread. There is the possibility that these compounds may affect the hypothalamic–pituitary–adrenal (HPA) axis and that this may explain their activity in some in vivo models. There is tremendous precedent for the actions of estrogens on the HPA axis. For example, sex differences in stress responses exist at both the behavioral and biochemical levels and across species (Bowman et al. 2002; Walf and Frye 2005; Kajantie and Phillips 2006). Estrogens affect a variety of rodent behavioral models of anxiety, although the effects vary with the model, dosing regimen and dose. These conflicting data likely result from the system's inherent complexity, but may also be influenced by the inherently different activity of ERα and ERβ and the fact that the best-studied estrogen, 17β-estradiol, can activate both ER subtypes.

ERβ has the potential to modulate the HPA axis in that it is expressed in the adrenal gland of a variety of species (Saunders et al. 1997; Albrecht et al. 1999) and is the dominant ER in the rat paraventricular nucleus (Shughrue et al. 1997). Moreover, one of the models where ERB-041 and WAY-202196 have profound effects is the Lewis rat adjuvant-induced arthritis model (Harris et al. 2003; Mewshaw et al. 2005; Follettie et al. 2006). Lewis rats are hypersensitive to inflammatory stimuli because this strain does not secrete appropriate corticotropin releasing hormone from the paraventricular nucleus. In fact, these rats have reduced basal ERα and ERβ expression in this nucleus (Tonelli et al. 2002). Finally, recent studies have begun to implicate ERβ in influencing the function of the HPA axis (Isgor et al. 2003; Miller et al. 2004; Lund et al. 2005, 2006).

We have examined ERβ mRNA and protein levels in a rat model of immobilization stress. Immobilization is a very potent stressor and leads to a large activation of the HPA axis as well as sympathoadrenal catecholaminergic systems. Pretreatment with estradiol benzoate has been previously shown to blunt stress-induced plasma ACTH and to modulate a variety of basal and stress-induced changes in gene expression in central and peripheral catecholaminergic locations (Serova et al. 2005). Repeated immobilization stress was found to dramatically upregulate ERβ in the adrenal medulla at both the mRNA and protein levels (Sabban E, unpublished observations). Interestingly, on a Western blot, several immunoreactive species are detected, with the predominant one being at 45 kD. The characterization of this isoform's sequence remains to be determined. Studies are underway to determine the effect of WAY-200070, compared to estradiol benzoate, on a variety of immobilization stress-induced biochemical responses.

6 Summary and Future Directions

The preclinical biology of selective ERβ agonists is as impressive as it is diverse. Using a set of selective agonists, we have discovered that manipulation of this receptor's activity may have value in treating human diseases such as inflammatory bowel disease, rheumatoid arthritis, endometriosis, and sepsis. In fact, phase II clinical trials are currently underway with ERB-041 for rheumatoid arthritis and endometriosis. Although there were hints that nonsubtype selective estrogens might affect some of these diseases (but not endometriosis), their potential beneficial effects were overshadowed by their effects on classic target tissues such as the uterus and mammary gland. Despite this progress, however, several key questions remain about ERβ selective agonists: first, and foremost from a scientific perspective, what are the molecular mechanism(s) behind the in vivo activities observed? Second, do these series of compounds reveal all aspects of ERβ biology or will other selective compounds have a different spectrum of activity? Lastly, and most important from a pharmaceutical perspective, what is the long-term safety profile of these compounds, and will their preclinical profile be mirrored in the human analog of these diseases? The next year or so will provide answers to some of these questions.

Acknowledgements. Our accomplishments in the field of ERβ research were enabled by the cooperative efforts of a number of scientists. First and foremost, Chris Miller, Mike Malamas, Rick Mewshaw, and Eric Manas led the chemistry effort to design and deliver highly selective ERβ agonists for biological characterization. Among the biologists, Jim Keith made key discoveries of in vivo activity; C. Richard Lyttle supported the team and suggested the endometriosis model evaluation. Several external collaborators have also contributed their expertise: Kaylon Bruner-Tran and Kevin Osteen (Vanderbilt University; endometriosis model), Kim Barrett and Silvia Resta-Lenert (University of California, San Diego; MDR1aKO mouse), Esther Sabban and Lydia Serova (New York Medical College, immobilization stress model), Steve Opal (Brown University; sepsis models). Finally I thank Rick Winneker for valuable suggestions on this manuscript.

References

Albrecht ED, Babischkin JS, Davies WA, Leavitt MG, Pepe GJ (1999) Identification and developmental expression of the estrogen receptor alpha and beta in the baboon fetal adrenal gland. Endocrinology 140:5953–5961

Angele MK, Schwacha MG, Ayala A, Chaudry IH (2000) Effect of gender and sex hormones on immune responses following shock. Shock 14:81–90

Bowman RE, Ferguson D, Luine VN (2002) Effects of chronic restraint stress and estradiol on open field activity, spatial memory, and monoaminergic neurotransmitters in ovariectomized rats. Neuroscience 113:401–410

Bruner KL, Matrisian LM, Rodgers WH, Gorstein F, Osteen KG (1997) Suppression of matrix metalloproteinases inhibits establishment of ectopic lesions by human endometrium in nude mice. J Clin Invest 99:2851–2857

Buras JA, Holzmann B, Sitkovsky M (2005) Animal models of sepsis: setting the stage. Nat Rev Drug Discov 4:854–865

Crabtree JS, Zhang X, Peano BJ, Zhang Z, Winneker RC, Harris HA (2006) Development of a mouse model of mammary gland versus uterine tissue selectivity using estrogen- and progesterone-regulated gene markers. J Steroid Biochem Mol Biol 101:11–21

Cristofaro PA, Opal SM, Palardy JE, Parejo NA, Jhung J, Keith JC Jr, Harris HA (2006) WAY-202196, a selective estrogen receptor-beta agonist, protects against death in experimental septic shock. Crit Care Med 34:2188–2193

Follettie MT, Pinard M, Keith JC, Wang LL, Chelsky D, Hayward C, Kearney P, Thibault P, Paramithiotis E, Dorner AJ, Harris HA (2006) Organ messenger ribonucleic acid and plasma proteome changes in the adjuvant-induced arthritis model: responses to disease induction and therapy with the estrogen receptor-beta selective agonist ERB-041. Endocrinology 147:714–723

Grummer R, Schwarzer F, Bainczyk K, Hess-Stumpp H, Regidor P, Schindler A, Winterhager E (2001) Peritoneal endometriosis: validation of an in vivo model. Human Reprod 16:1736–1743

Harris HA (2007) Estrogen receptor-beta: recent lessons from in vivo studies. Mol Endocrinol, pp 2111–2113

Harris HA, Katzenellenbogen JA, Katzenellenbogen BS (2002) Characterization of the biological roles of the estrogen receptors ER alpha and ER beta, in estrogen target tissues in vivo through the use of an ER alpha-selective ligand. Endocrinology 143:4172–4177

Harris HA, Albert LM, Leathurby Y, Malamas MS, Mewshaw RE, Miller CP, Kharode YP, Marzolf J, Komm BS, Winneker RC, Frail DE, Henderson RA, Zhu Y, Keith JC Jr (2003) Evaluation of an estrogen receptor-beta agonist in animal models of human disease. Endocrinology 144:4241–4249

Harris HA, Bruner-Tran KL, Zhang X, Osteen KG, Lyttle CR (2005) A selective estrogen receptor-beta agonist causes lesion regression in an experimentally induced model of endometriosis. Human Reproduction 20:936–941

Hillisch A, Peters O, Kosemund D, Muller G, Walter A, Schneider B, Reddersen G, Elger W, Fritzmeier KH (2004) Dissecting physiological roles of estrogen receptor alpha and beta with potent selective ligands from structure-based design. Mol Endocrinol 18:1599–1609

Homma H, Hoy E, Xu D-Z, Lu Q, Feinman R, Deitch EA (2005) The female intestine is more resistant than the male intestine to gut injury and inflammation when subjected to conditions associated with shock states. Am J Physiol Gastrointest Liver Physiol 288:G466–G472

Isgor C, Cecchi M, Kabbaj M, Akil H, Watson SJ (2003) Estrogen receptor beta in the paraventricular nucleus of hypothalamus regulates the neuroendocrine response to stress and is regulated by corticosterone. Neuroscience 121:837–845

Kajantie E, Phillips DIW (2006) The effects of sex and hormonal status on the physiological response to acute psychosocial stress. Psychoneuroendocrinology 31:151–178

Leventhal L, Brandt MR, Cummons TA, Piesla MJ, Rogers KE, Harris KA. An estrogen receptor-β agonist is active in models of inflammatory and chemical-induced pain. European Journal of Pharmacology 553:146–148

Lund TD, Rovis T, Chung WCJ, Handa RJ (2005) Novel actions of estrogen receptor-beta on anxiety-related behaviors. Endocrinology 146:797–807

Lund TD, Hinds LR, Handa RJ (2006) The androgen 5 alpha-dihydrotestosterone and its metabolite 5 alpha-androstan-3 beta,17 beta-diol inhibit the hypothalamo-pituitary-adrenal response to stress by acting through estrogen receptor beta-expressing neurons in the hypothalamus. J Neurosci 26:1448–1456

Malamas MS, Manas ES, McDevitt RE, Gunawan I, Xu ZB, Collini MD, Miller CP, Dinh T, Henderson RA, Keith JC Jr, Harris HA (2004) Design and synthesis of aryl diphenolic azoles as potent and selective estrogen receptor-beta ligands. J Med Chem 47:5021–5040

Mewshaw RE, Edsall RJ, Yang CJ, Manas ES, Xu ZB, Henderson RA, Keith JC, Harris HA (2005) ER beta ligands. 3. Exploiting two binding orientations of the 2-phenylnaphthalene scaffold to achieve ER beta selectivity. J Med Chem 48:3953–3979

Miller CP (2002) SERMs: evolutionary chemistry, revolutionary biology. Curr Pharmaceut Des 8:2089–2111

Miller WJS, Suzuki S, Miller LK, Handa R, Uht RM (2004) Estrogen receptor (ER)beta isoforms rather than ER alpha regulate corticotropin-releasing hormone promoter activity through an alternate pathway. J Neurosci 24:10628–10635

Resta-Lenert S, Smitham J, Barrett KE (2005) Epithelial dysfunction associated with the development of colitis in conventionally housed mdr1a(-/-) mice. Am J Physiol Gastrointest Liver Physiol 289:G153–G162

Saunders PTK, Maguire SM, Gaughan J, Millar MR (1997) Expression of oestrogen receptor beta (ER-beta) in multiple rat tissues visualised by immunohistochemistry. J Endocrinol 154:R13–R16

Serova LI, Maharjan S, Sabban EL (2005) Estrogen modifies stress response of catecholamine biosynthetic enzyme genes and cardiovascular system in ovariectomized female rats. Neuroscience 132:249–259

Shughrue PJ, Lane MV, Merchenthaler I (1997) Comparative distribution of estrogen receptor-alpha and -beta mRNA in the rat central nervous system. J Comp Neurol 388:507–525

Tonelli L, Kramer P, Webster JI, Wray S, Listwak S, Sternberg E (2002) Lipopolysaccharide-induced oestrogen receptor regulation in the paraventricular hypothalamic nucleus of Lewis and Fischer rats. J Neuroendocrinol 14:847–852

Veeneman GH (2005) Non-steroidal subtype selective estrogens. Curr Med Chem 12:1077–1136

Vincent J-L, Abraham E (2005) The last 100 years of sepsis. Am J Respir Crit Care Med 173:256–263

Walf AA, Frye CA (2005) Antianxiety and antidepressive behavior produced by physiological estradiol regimen may be modulated by hypothalamic-pituitary-adrenal axis activity. Neuropsychopharmacology 30:1288–1301

ns, Vol. 1, pp. 163–180

Exploiting Nongenomic Estrogen Receptor-Mediated Signaling for the Development of Pathway-Selective Estrogen Receptor Ligands

C. Otto[✉], S. Wessler, K.-H. Fritzemeier

TRG Gynecology and Andrology, Bayer Schering Pharma AG, Müllerstr. 178, 13342 Berlin, Germany
email: *christiane.otto@schering.de*

1	Introduction	164
2	Establishment of In Vitro Assays Monitoring Genomic and Nongenomic ERα-Mediated Actions	166
3	Identification and In Vitro Characterization of Tool Compounds	168
4	Testing of Tool Compounds In Vivo	172
4.1	Protection from Ovariectomy-Induced Bone Loss	173
4.2	Uterine Growth Assays	174
4.3	Mammary Gland Whole Mount Assays	178
5	Outlook	178
References		179

Abstract. Different molecular mechanisms mediate the diverse biological effects of estrogens. The classical genomic mechanism is based on the function of the ER as a ligand-dependent transcription factor that binds to estrogen-response elements (EREs) in promoters of target genes to initiate gene expression. These direct genomic effects play a prominent role in the regulation of reproductive function. In contrast, nongenomic effects mediated by the classical ER have been demonstrated to activate PI3K, leading to the activation of endothelial NOS (eNOS) and hence vasorelaxation. Pathway-selective ER ligands might represent a novel option for hormone replacement therapy. Here we describe the identification and in vitro characterization of tool compounds that bind the ER reasonably well but exhibit low transcriptional activity on ERE-driven promoters. However, these compounds behave as potent stimulators of PI3K/Akt activation in vitro and lead to aortic vessel relaxation, a mechanism that is thought to be driven by nongenomic ER action. In a second set of experiments, we analyze how the in vitro pathway selectivity translates into the in vivo situation. We examine our tool compounds in comparison to estradiol and estren in the following paradigms: bone protection, uterine growth assays, and mammary gland assays.

1 Introduction

Estradiol regulates a variety of physiological processes such as reproduction (Hart and Davie 2002), neuroprotection (Garcia-Segura et al. 2001), the cardiovascular system (Medina et al. 2003), and bone turnover (Kousteni et al. 2002).

During the last few years, it has become evident that different molecular mechanisms contribute to these diverse biological effects of estrogen. The classical genomic effects rely on the function of the estrogen receptor (ER) as a ligand-dependent transcription factor binding to estrogen response elements (EREs) in promoters of target genes. These genomic effects are critically involved in the establishment and maintenance of reproductive function (Hart and Davie 2002). In contrast, rapid nongenomic effects are initiated at the plasma membrane and lead to an activation of cytoplasmic signal transduction cascades. These nongenomic effects have been suggested to mediate the effects of estrogen in bone (Kousteni et al. 2002) or the cardiovascular system (Simoncini et al. 2004). In bone, estrogen has anti-apoptotic effects on osteoblasts and osteocytes. These effects are mediated via nongenomic activation of the Src/Shc/ERK pathway and repression of JNK signaling cascades, leading to modulation of the activity of transcription factors such as Elk-1, CREB, and c-Fos/c-Jun (Kousteni et al. 2001). In endothelial cells, ERα/PI3K interaction leads to rapid stimulation of eNOS activity followed by NO production and vasorelaxation (Simoncini et al. 2002).

In postmenopausal women, estrogens have beneficial effects on bone and the vascular system. It has been suggested that these beneficial effects are partly mediated via nongenomic effects. In contrast, the stimulation of uterine epithelial cell proliferation – an unwanted side effect of estrogen-only therapy – depends on binding of the ER to EREs and thus genomic effects. Mice harboring point mutations within the DNA-binding region of the ER (Jakacka et al. 2002) do not show uterine growth in response to estradiol and therefore underline the importance of genomic ER signaling for uterine growth.

These findings may support the conclusion that pathway-selective ER ligands leading to stimulation of nongenomic ER responses, but showing only reduced genomic effects may be of value for novel hormone replacement regimens. A few years ago, estren (4-estren-3α,17β-diol) was described as a pathway-selective ER ligand that prevents ovariectomy-induced bone loss via its nongenomic effects, whereas it does not have any stimulatory action on the uterus due to its reduced genomic activity (Kousteni et al. 2002). However, a few years later it became evident that estren acts as a potent androgen in vivo that increases the levator ani muscle and seminal vesicle weights in male orchidectomized mice (Krishan et al. 2005) and that promotes androgen phenotypes in primary lymphoid organs that are clearly independent of the presence of ERs (Islander et al. 2005).

Strong androgenic compounds, however, are not of interest for hormone replacement therapy in women. In order to identify pathway-selective estrogens, we addressed two key questions:

1. Do pathway-selective estrogens without androgenic activity exist?
2. How does the pathway selectivity observed in vitro translate into the in vivo situation?

For that purpose we developed in vitro assays monitoring genomic and nongenomic effects mediated by the classical ERα. Tool compounds identified with the desired in vitro profile were then subjected to different in vivo assays, i.e., protection from ovariectomy-induced bone loss, uterine growth assays, and mammary gland whole mount assays. For reference, estren and 17β-estradiol were included.

2 Establishment of In Vitro Assays Monitoring Genomic and Nongenomic ERα-Mediated Actions

To monitor genomic activity of different ER ligands, we performed transactivation assays using the osteosarcoma cell line U2-OS transiently transfected with human ERα and different luciferase reporter constructs, i.e., pBL(ERE)$_2$tkLuc$^+$ (Hillisch et al. 2004) or pC3-LUC$^+$ (Tzukerman et al. 1994). From our previous studies, it was already established that estrogenic potency measured in these transactivation assays reflected uterotrophic activity in vivo quite well (Hillisch et al. 2004).

For the analysis of nongenomic effects, we exploited two different approaches: rapid estradiol-induced phosphorylation of Akt and ERK1/2 as well as estradiol-dependent interaction of ER and Src.

Fig. 1a–d. Establishment of assays testing nongenomic activities mediated by the classic ERα. **a** U2OS cells were transfected with the ERα expressing Hego plasmid or empty vector (pSG5) and treated for 30 min with 1 nM E2 or vehicle. Activated Akt and ERK1/2 were detected with phosphospecific antibodies. Immunoblots were reprobed with a non-phosphospecific Akt antibody for loading control and with an ERα-specific antibody to check for proper expression after transient transfection. **b** U2OS cells were transiently transfected with ERα and pretreated for 30 min with 20 µM of different inhibitors followed by stimulation with 1 nM E2 for 30 min. Activation of Akt and ERK1/2 was assessed using phosphospecific antibodies, blots were reprobed for loading control with non-phosphospecific ERK antibodies. **c** Coimmunoprecipitation of ERα and Src from lysates of MCF-7 cells stimulated for 30 min with 1 nM E2. Immunoprecipitation was performed with ERα-specific antibodies, interacting partners were detected by Western blot. **d** GST pulldown experiments using lysates from T47D cells stimulated with 10 nM E2 for 1 h and different Src-GST fusion proteins. The *upper panel* shows the amount of interacting ERα, the *middle panel* shows the input control of ERα, and the *bottom panel* shows the loading control for the GST fusion proteins. GST-metSH2 encompasses the first ATG of Src, the SH3 as well as the SH2 domain. GST-SH3 contains the SH3 domain, and GST-SH2 contains the SH2 domain of Src. ERα interacts in a hormone-dependent manner with the SH2 domain of Src. (Data depicted in Figs. 1–4 and shown in Table 1 are reprinted from Wessler et al. 2006, with permission from Elsevier)

Pathway-Selective Estrogen Receptor Ligands

In a first set of experiments, we transiently transfected U2-OS cells that are devoid of endogenous ER with either an expression vector encoding human ERα (Hego) or the empty vector pSG5. Cells were then treated for 30 min with 1 nM E2 and activation of Akt and ERK1/2 was assessed by Western blot using phosphospecific antibodies. As depicted in Fig. 1a, phosphorylation and hence activation of both Akt

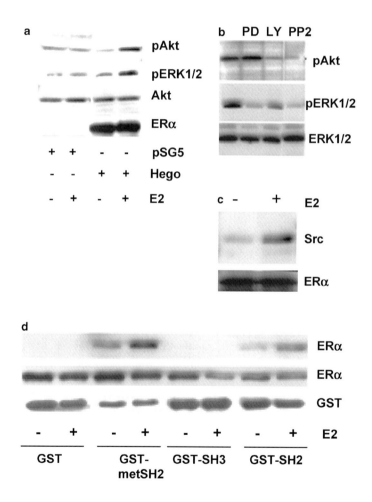

and ERK1/2 were clearly ER- and estradiol-dependent. These findings are in line with previous observations suggesting that in breast cancer cells nongenomic activation of both Src and PI3K by estradiol is important for cellular proliferation (Castoria et al. 2001). To investigate proximal signaling events involved in Src and ERK1/2 activation, we pretreated the transfected cells with different kinase inhibitors before stimulation with estradiol. As expected, pretreatment with the MEK inhibitor PD98059 completely blunted ERK1/2 phosphorylation but had no effect on Akt activation. The PI3K inhibitor LY294002 and the Src kinase inhibitor PP2 impaired both Akt and ERK1/2 activation, indicating that PI3K as well as Src are a prerequisite for ERK1/2 activation (Fig. 1b). In coimmunoprecipitation experiments using cell lysates from transfected U2-OS cells, we were able to demonstrate that Src and ERα interacted in an estradiol-dependent manner (Fig. 1c).

Src contains two protein–protein interaction domains: the SH2 and SH3 domain. To analyze which of both domains is of importance for ER/Src interaction, we constructed GST-fusion proteins spanning different regions of Src. In GST-pulldown experiments, we observed strong estradiol-dependent interaction between ER and the GST-fusion protein encompassing the first 247 aa of Src and thus the SH2 and SH3 domain (GST-metSH2). Using GST-Src fusion constructs encoding either the SH2 or SH3 domain, we were able to refine the ER interaction site of Src to the SH2 domain (Fig. 1d). There was no unspecific interaction between the ER and the GST protein, indicating that the observed interaction between ER and the SH2 domain of Src was specific. This result is in line with previous findings (Migliaccio et al. 2000).

In summary, we succeeded in establishing two different readout paradigms to assess nongenomic activities of ER ligands in vitro: estradiol-dependent ER/Src interaction and rapid activation of Akt and ERK1/2 in transfected U2OS cells.

3 Identification and In Vitro Characterization of Tool Compounds

To identify tool compounds that could serve as templates for new pathway-selective ER ligands, we have chosen molecules from Schering

AG's compound library that were characterized by strong ER binding but exhibited only very low stimulation of ERE-dependent transcription. Structures of selected tool compounds are depicted in Fig. 2, whereas the binding and transactivation properties are shown in Table 1. Estren was included for comparison. Like estradiol, compounds A and B exhibit high binding affinity for ERα, but they are almost inactive in the activation of ERE-driven target gene induction. In contrast, estren shows already strongly diminished binding affinity for ERα that translates into weak ER-dependent transcriptional activity (Table 1). When we analyzed androgen-receptor-dependent activation of luciferase gene expression, estren was almost as potent as the reference metribolone, whereas compounds A and B were completely devoid of androgenic activity. Therefore, compounds A and B, but not estren, most likely reflect the desired in vitro profile of a pathway-selective estrogen. To test this, we had to examine the nongenomic properties of both compounds. We demonstrated that estren (being only slightly less potent) and both test compounds were able to induce rapid induction of Akt and ERK1/2 activation in serum-starved MCF7 cells (Wessler et al. 2006). In addition, all compounds were able to specifically stimulate ER/Src interaction (Fig. 3).

To further elucidate the nongenomic activities of our test compounds in a more physiological context, we assayed nongenomic activation of

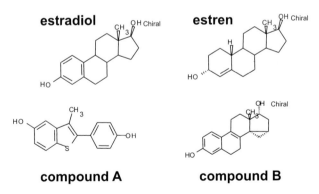

Fig. 2. Structures of selected tool compounds for in vitro testing

Table 1 In vitro characteristics of selected tool compounds

Compound	CF* ERα binding	CF ERα transactivation	CF AR transactivation	Nongenomic activity
Estradiol	1	1	No effect	++
A	0.7	250	No effect	+++
B	2.9	1000	No effect	++
Estren	45	1000	1.5	++

*CF = EC50 compound/EC50 reference;
reference = estradiol in the case of ER and metribolone in the case of AR

eNOS by estradiol, leading to vessel relaxation (Simoncini et al. 2004). Phenylephrine-precontracted rat aortic rings were exposed to cumulative doses of either E2, compound A, compound B, or estren. Among

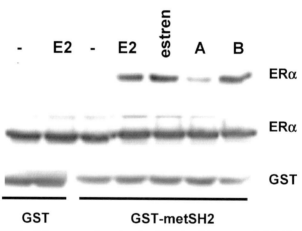

Fig. 3. GST pulldown experiments using T47D cell lysates and GST-Src fusion protein. Estradiol and the selected tool compounds, but not vehicle, stimulate direct interaction of ERα with Src. The *upper panel* shows the amount of interacting ERα, the *middle panel* shows the input control for ERα, and the *bottom panel* shows the amount of GST-Src fusion protein loaded onto gel. GST alone does not interact with ERα and thus serves as negative control

all compounds analyzed, estren exhibited the lowest potency and efficiency for the induction of vasorelaxation. While contraction of aortic rings was reduced to 50% of control levels at 5 µM E2 or substance A or at 3 µM substance B, 15 µM estren was required to evoke the same response (Fig. 4). Substance A was the most efficient compound with

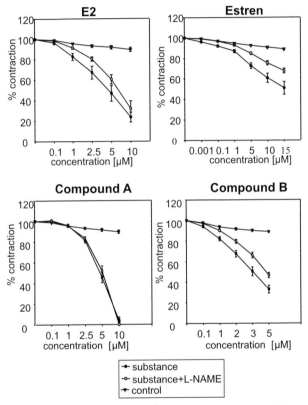

Fig. 4. Vasorelaxation of rat aortic rings induced by estradiol and the test compounds. Phenylephrine-precontracted aortic rings were treated with cumulative doses of compounds (*closed circles*) or vehicle (*closed triangles*) and relaxation was measured. Selected rings were pretreated with 1 mM L-NAME (*open circles*). Graphs show contraction expressed as a percentage of maximal phenylephrine-induced vasoconstriction

respect to the induction of vasorelaxation. At 10 µM, contraction of aortic rings was completely abolished by compound A. Pretreatment of aortic rings with 1 mM L-NAME significantly reduced vasodilation induced by estradiol, estren, and substance B, indicating an involvement of eNOS activation. In contrast, vasodilation induced by compound A was insensitive toward L-NAME, indicating that compound A-mediated vasorelaxation may involve an eNOS-independent mechanism (Fig. 4). To test this hypothesis, we analyzed rapid nongenomic stimulation of eNOS activity in cultured HUVECs (human umbilical vein endothelial cells). As expected from the vasorelaxation experiments, compound A was not able to stimulate eNOS activity in HUVECs, whereas all other tested substances led to rapid eNOS activation (T. Simoncini, unpublished observations). Most likely, compound A provokes rapid vasorelaxation by direct action on the smooth vasculature. It has been demonstrated that rapid nongenomic ER effects on the vasculature can also lead to eNOS-independent vasorelaxation. For example, whole-cell patch-clamp analysis has demonstrated that estradiol directly inhibits calcium channel activity in the plasma membrane of vascular smooth muscle cells, leading to reduced intracellular calcium levels and hence vasorelaxation (Salom et al. 2002).

Taken together, our results clearly demonstrate that compounds A and B show the desired in vitro profile of a pathway-selective estrogen, i.e., they are potent ER ligands devoid of androgenic activity that stimulate nongenomic ER-mediated effects but exhibit only negligible transcriptional activity.

4 Testing of Tool Compounds In Vivo

Several substances from the Schering library were identified that exhibited a similar in vitro profile as compounds A and B. For in depth in vivo characterization of such compounds, we chose compound A (Fig. 2) and the nonsteroidal compound C (structure to be disclosed). Reference compounds were estradiol and estren. We performed the following assays:

1. Protection from ovariectomy-induced bone loss
2. Uterine growth assays
3. Mammary gland whole mount assays.

The aim was to analyze how the relative in vitro pathway selectivity is translated into the in vivo situation. We were interested to see whether the reduced genomic effects of compounds A and C led to a stronger dissociation of bone vs uterus and mammary gland effects if compared to the reference estradiol, which by definition does not show any dissociation.

4.1 Protection from Ovariectomy-Induced Bone Loss

The experiments were performed in close analogy to previous studies (Kousteni et al. 2002). We either ovariectomized or sham operated 6-month-old Swiss Webster mice and treated them for 6 weeks subcutaneously with vehicle, 6.4 µg/kg E2 or different doses of estren, compounds A and C. After 6 weeks of treatment cortical bone area, trabecular bone mass and relative uterine weight were assessed. The results are summarized in Table 2. Estradiol at 6.4 µg/kg completely protected ovariectomized mice from bone loss. Upon estradiol treatment, trabecular as well as cortical bone density were as high as in sham-operated control animals. At its bone-protective dose, estradiol showed full uterotrophic activity, i.e., relative uterine weight was as high as in sham-operated animals. These results, i.e., no dissociation between estradiol-stimulated uterine and bone effects, are in line with the observation that estradiol activates genomic and nongenomic effects. Like estradiol, estren proved to be fully effective with regard to protection from trabecular and cortical bone loss. However, a much higher dose, 5 mg/kg, was required to achieve this. Contrary to published results (Kousteni et al. 2002), we observed full uterotrophic activity of estren at its bone-protective dose (Table 2). Whereas estren had positive effects on trabecular as well as cortical bone, compound A at 0.1 mg/kg was only effective on cortical bone and compound C at 1 mg/kg protected animals only from trabecular bone loss. There was only a very slight increase of relative uterine weight that was significantly different from vehicle treatment. From that data set, one may conclude that compounds A and C in contrast to estren and estradiol display a dissociated pattern of action, i.e., full protection from either cortical or trabecular bone loss, but almost no stimulation of uterine growth. However, the interpretation of these results is much more complicated. It has been

Table 2 Protection from ovariectomy-induced bone loss

Compound	Bone-protective dose	Effect on trabecular bone	Effect on cortical bone	Relative uterine weight
Estradiol	6.4 µg/kg	+++	+++	+++
Compound A	0.1 mg/kg	No effect	+++	(+)
Compound C	1 mg/kg	+++	No effect	(+)
Estren	5 mg/kg	+++	+++	+++

+++ Fully effective, i.e., no significant difference to sham operated animals; (+) extremely little effect compared to vehicle treatment

demonstrated by others and by our laboratory that estren is a relatively strong androgen (Islander et al. 2005; Krishan et al. 2005). The increase in relative uterine weight that we observe not only in mice but also in rats (data not shown) is most likely due to androgenic effects of estren on the myometrium and not related to an increase in endometrial epithelial cell number typically caused by ER ligands.

4.2 Uterine Growth Assays

To further elucidate the dissociation of bone vs uterine effects of our test compounds, we performed several uterine growth assays using different treatment times and different mouse strains. We analyzed different readout parameters such as relative uterine weight, epithelial cell height, BrdU incorporation, and target gene induction. Again estradiol and estren served as reference compounds.

In a first set of experiments, we examined uterine growth in either juvenile B6D2F1 mice or in 3-month-old Swiss Webster mice. Two weeks after ovariectomy, mice were treated with vehicle or different subcutaneous doses of compounds for 3 days. On day 4, relative uterine weight as well as epithelial cell height in the endometrium were determined. The results obtained from juvenile B6D2F1 mice and older Swiss Webster mice were in good accordance. In contrast to the results observed in uteri after 6 weeks of treatment, relative uterine weight gain at the bone-protective doses was roughly the same for compounds A and C and for estren but reached only 20%–30% of the effect on uterine weight found after E2 treatment (data not shown). Thus estren as well as our tool compounds seem to show a clear dissociation between bone and uterine effects. We also analyzed epithelial cell height in the uteri after 3 days of treatment. Results are depicted in Fig. 5. As expected, E2 led to maximal stimulation of epithelial cell height. Estren and compound C at their bone-protective doses showed only a partial effect, with compound C being more effective than estren (Fig. 5). However, compound A at its bone-protective dose of 0.1 mg/kg did not show any significant effect on epithelial cell height if compared with vehicle treatment.

To determine the degree of dissociation of bone vs uterine effects, we divided the relative efficacy of the compounds at their bone-protective dose for bone effects by the relative efficacy for uterine effects at the same dose. The efficacy of the reference estradiol was set to 100%. A dissociation factor of 1 is obtained by definition for estradiol. Es-

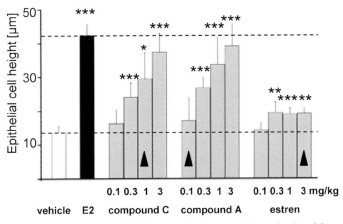

Fig. 5. Three-month-old Swiss Webster mice were ovariectomized and 2 weeks after ovariectomy mice were treated subcutaneously for 3 days. Autopsy was performed on day 4 and uterine epithelial cell height was analyzed. *Triangles within the bars* mark the bone-protective dose of each compound. *Stars* indicate significant differences vs vehicle treatment. Estradiol at its bone-protective dose shows full uterotrophic activity; compound A at its bone-protective dose has no significant effect on epithelial cell height. Compound C and estren at their bone-protective dose show only partial effects on epithelial cell height

tren and compound A showed dissociation factors of approximately 5, whereas compound C was less dissociated.

To more closely investigate the uterine effects of the tool compounds, we performed short-term uterine growth assays (compound administration for only 18 h in ovariectomized mice) using BrdU incorporation in epithelial cells and target gene induction as readout parameters. In these short-term experiments, estradiol at its bone-protective dose was again fully active with regard to BrdU incorporation in uterine epithelial cells (data not shown), lactotransferrin, and cyclin E1 mRNA induction (Fig. 6). Estren and compounds A and C at their bone-protective dose do not induce lactotransferrin mRNA above vehicle levels (Fig. 6). This result fits with the observation that the lactotransferrin promoter carries EREs (Liu and Teng 1992). Therefore it is expected that compounds with strongly reduced genomic activity fail to induce this gene at doses that are effective for the prevention of ovariectomy-induced bone loss – a paradigm that is thought to depend partly on nongenomic mechanisms (Kousteni et al. 2002). With regard to cyclin E1 induction, all three compounds exhibit partial agonism, but the androgenic compound estren shows only a minor effect (Fig. 6).

A first conclusion from these studies may be that the mechanism of cyclin E1 induction is different from the induction of lactotransferrin and may partly rely on nongenomic mechanisms. Cyclin E1 is a S-phase cyclin and therefore might be suitable to predict the extent of BrdU incorporation and thus S-phase entry in uterine epithelial cells. This is indeed the case, since all three test compounds lead to stimulation of uterine epithelial cell proliferation at their bone-protective dose. Whereas compounds A and C are as effective as estradiol, estren is only half as effective (data not shown).

The results obtained so far support the conclusion that compared to estradiol ER ligands with reduced genomic activity show a superior dissociation of bone vs uterine effects in vivo.

In order to investigate the uterine effects of the tool compounds after long-term exposure, we also designed experiments that allow the analysis of cellular proliferation and target gene induction after 3 weeks of compound treatment. The evaluation of these experiments is in progress.

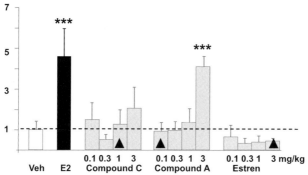
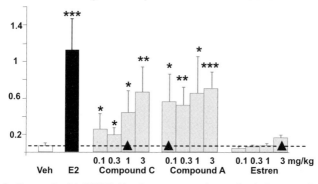

Fig. 6. Six-week-old C57BL/6 mice were ovariectomized. Two weeks after ovariectomy, animals were injected with different doses of compounds subcutaneously. Mice were killed 20 h after compound administration. RNA was prepared from collected uteri and expression of lactotransferrin and cyclin E1 was analyzed. *Triangles* indicate the bone-protective dose of each compound. According to their low genomic activity, compounds A and C and estren do not induce the ERE-dependent gene lactotransferrin at their bone-protective doses, whereas estradiol shows full activity as expected. With regard to cyclin E1 induction, estradiol shows full activity, compounds A and C have only partial effects, and estren does not show any effect that is significantly different from vehicle treatment

4.3 Mammary Gland Whole Mount Assays

To analyze the dissociation of bone vs mammary gland effects, we ovariectomized 6-week-old C57BL/6 mice. Two weeks after ovariectomy, animals were treated subcutaneously with different doses of estradiol and the three test compounds for 3 weeks. Endbud formation, BrdU incorporation in mammary epithelial cells, and target gene induction were analyzed. Since analysis of the two latter items is still under investigation, we focus here on endbud formation.

While estradiol stimulates endbud formation with a relative efficacy of 100% and an ED50 of 0.27 µg/kg per day, estren as an androgenic compound is almost completely ineffective and shows only an efficacy of about 5%. Compound A (relative efficacy 72%, ED50 = 0.09 mg/kg per day) and compound C (relative efficacy 60%, ED50 = 0.3 mg/kg per day) are clearly less potent and effective than estradiol. To evaluate the dissociation of bone vs mammary gland effects, we again determined dissociation factors. At the bone-protective dose, the relative efficacy of the compounds in the bone assay were divided by the relative efficacy in the mammary gland assay. In comparison to estradiol (dissociation factor 1), compounds A and C show a much better dissociation (dissociation factor 2), i.e., the compound dose that is required for 100% bone protection leads only to 50% efficacy with regard to endbud formation in the mammary gland. As expected from its androgenic nature, estren shows the greatest dissociation of bone vs mammary gland effects (dissociation factor 15).

5 Outlook

The aim of the experiments highlighted above was to investigate whether pathway-selective ER ligands exist and how this pathway-selectivity observed in vitro would translate into the in vivo situation.

In summary, we were able to identify several compounds with the desired in vitro profile, i.e., high-affinity ER ligands that were characterized by reduced stimulation of genomic effects and strong activation of nongenomic ER effects. In contrast to estren, these compounds were devoid of any androgenic activity. Although analysis of long-term in vivo experiments is still ongoing, our current results already demonstrate that compounds with reduced genomic activity in vitro show a superior in vivo profile in comparison to the undissociated ER ligand estradiol. Whereas estradiol shows full uterotrophic activity or is fully active in the mammary gland at doses that protect from ovariectomy-induced bone loss, our tool compounds at their bone-protective doses show clearly reduced efficacy with respect to their effects on the mammary gland and the uterus. Compounds exhibiting such an in vivo profile maybe promising candidates for novel hormone replacement strategies in postmenopausal women that may lead to less stimulatory effects in the mammary gland and the uterus.

However, highly differentiating profiling of ER ligands and the development of in vitro assays that allow faithful prediction of the in vivo activity will be necessary in order to screen for such ligands in a high throughput scale.

References

Castoria G, Migliaccio A, Bilancio A, Di Domenico M, de Falco A, Lombardi M, Fiorentino R, Varicchio L, Barone MV, Auricchio F (2001) PI3-kinase in concert with Src promotes the S-phase entry of estradiol-stimulated MCF-7 cells. EMBO J 20:6050–6059

Garcia-Segura LM, Azcoitia I, Don Carlos LL (2001) Neuroprotection by estradiol. Prog Neurobiol 63:29–60

Hart LL, Davie RR (2002) The estrogen receptor: more than the average transcription factor. Biochem Cell Biol 80:335–341

Hillisch A, Peters O, Kosemund D, Muller G, Walter A, Schneider B, Reddersen G, Elger W, Fritzemeier KH (2004) Dissecting physiological roles of estrogen receptor alpha and beta with potent selective ligands from structure-based design. Mol Endocrinol 18:1599–1609

Islander U, Hasseus B, Erlandsson MC, Jochems C, Skrtic SM, Lindberg M, Gustafsson JA, Ohlsson C, Carlsten H (2005) Estren promotes androgen phenotypes in primary lymphoid organs and submandibular glands. BMC Immunol 6:16

Jakacka M, Ito M, Martinson F, Ishikawa T, Lee EJ, Jameson JL (2002) An estrogen receptor (ER)α deoxyribonucleic acid-binding domain knock-in mutation provides evidence for nonclassical ER pathway signalling in vivo. Mol Endocrinol 16:2188–2201

Kousteni S, Bellido T, Plotkin LI, O'Brien CA, Bodenner DL, Han L, Han K, DiGregorio GB, Katzenellenbogen JA, Katzenellenbogen BS, Roberson PK, Weinstein RS, Jilka RL, Manolagas SC (2001) Nongenotropic, sex-nonspecific signalling through the estrogen or androgen receptors: dissociation from transcriptional activity. Cell 104:719–730

Kousteni S, Chen JR, Bellido T, Han L, Ali AA, O'Brien CA, Plotkin L, Fu Q, Mancino AT, Wen Y, Vertino AM, Powers CC, Stewart SA, Ebert E, Parfitt AM, Weinstein RS, Jilka RL, Manolagas SC (2002) Reversal of bone loss in mice by nongenotropic signalling of sex steroids. Science 298:843–846

Krishan V, Bullock HA, Yaden BC, Liu M, Barr RJ, Montrose-Rafizadeh C, Chen K, Dodge JA, Bryant HU (2005) The nongenotropic synthetic ligand 4-estren-3alpha17beta-diol is a high-affinity genotropic androgen receptor agonist. Mol Pharmacol 67:744–748

Liu Y, Teng CT (1992) Estrogen response module of the mouse lactoferrin gene contains overlapping chicken ovalbumin upstream promoter transcription factor and estrogen receptor-binding elements. Mol Endocrinol 6:355–364

Medina RA, Aranda E, Verdugo C, Kato S, Owen GI (2003) The action of ovarian hormones in cardiovascular disease. Biol Res 36:325–341

Migliaccio A, Castoria G, Di Domenico M, de Falco A, Bilancio A, Lombardi M, Barone MV, Ametrano D, Zannini MS, Abbondanza C, Auricchio F (2000) Steroid-induced androgen receptor-oestradiol receptor beta-Src complex triggers prostate cancer cell proliferation. EMBO J 19:5406–5417

Norman AW, Litwack G (1987) Estrogens and progestins. In: Litwack G (ed) Hormones. Academic Press, London, pp 550–560

Salom JB, Burguete MC, Perez-Asensio FJ, Centeno JM, Torregrosa G, Alborch E (2002) Acute relaxant effects of 17-β-estradiol through non-genomic mechanisms in rabbit carotid artery. Steroids 67:339–346

Simoncini T, Fornari L, Mannella P, Varone G, Caruso A, Liao JK, Genazzani AR (2002) Novel non-transcriptional mechanisms for estrogen receptor signalling in the cardiovascular system. Interaction of estrogen receptor alpha with phosphatidyl 3-OH kinase. Steroids 67:935–939

Simoncini T, Mannella P, Fornari L, Caruso A, Varone G, Genazzani AR (2004) Genomic and non-genomic effects of estrogens on endothelial cells. Steroids 69:537–542

Tzukerman MT, Esty A, Santiso-Mere D, Danielian P, Parker MG, Stein RB, Pike JW, McDonnell DP (1994) Human estrogen receptor transactivational capacity is determined by both cellular and promoter context and mediated by two functionally distinct intramolecular regions. Mol Endocrinol 8(1):21–30

Wessler S, Otto C, Wilck N, Stangl V, Fritzemeier KH (2006) Identification of estrogen receptor ligands leading to activation of non-genomic signalling pathways while exhibiting only weak transcriptional activity. J Steroid Biochem Mol Biol 98:25–35

Ernst Schering Foundation Symposium Proceedings

Editors: Günter Stock
Monika Lessl

Vol. 1 (2006/1): Tissue-Specific Estrogen Action
Editors: K.S. Korach, T. Wintermantel

This series will be available on request from
Ernst Schering Research Foundation, 13342 Berlin, Germany

Printing: Krips bv, Meppel
Binding: Stürtz, Würzburg